17th Edition
IEE Wiring Regulations:
Design and Verification of Electrical
Installations

By the same author

17th Edition IEE Wiring Regulations: Inspection, Testing and Certification, ISBN 978-0-0809-6610-6
17th Edition IEE Wiring Regulations: Explained and Illustrated, ISBN 978-0-7506-8720-1
Electric Wiring: Domestic, ISBN 978-0-7506-8735-5
PAT: Portable Appliance Testing, ISBN 978-0-0809-6919-0
Wiring Systems and Fault Finding, ISBN 978-0-7506-8734-8
Electrical Installation Work, ISBN 978-0-7506-8733-1

17th Edition
IEE Wiring Regulations:
Design and Verification of Electrical Installations

Seventh Edition

Brian Scaddan, IEng, MIET

AMSTERDAM • BOSTON • HEIDELBERG • LONDON • NEW YORK
OXFORD • PARIS • SAN DIEGO • SAN FRANCISCO
SINGAPORE • SYDNEY • TOKYO
Newnes is an imprint of Elsevier

Newnes is an imprint of Elsevier
The Boulevard, Langford Lane, Kidlington, Oxford OX5 1GB, UK
225 Wyman Street, Waltham, MA 02451, USA

First edition 1995
Second edition 1999
Third edition 2001
Fourth edition 2002
Fifth edition 2005
Sixth edition 2008
Seventh edition 2011

British Library Cataloguing in Publication Data
A catalogue record for this book is available from the British Library

Library of Congress Cataloging-in-Publication Data
A catalog record for this book is availabe from the Library of Congress

ISBN: 978-0-08-096914-5   621.319240
                            435023

For information on all Newnes publications
visit our web site at books.elsevier.com

Printed and bound in Italy
11 12 13 14 15   10 9 8 7 6 5 4 3 2 1

In memory of
Ted Stocks
A friend and colleague

Contents

Preface

There are many electrical operatives who, quite innocently I am sure, select wiring systems based on the old adage of 'that's the way it's always been done' or 'we always use that size of cable for that circuit', etc. Unfortunately this approach, except for a few standard circuits, is quite wrong. Each wiring system should be designed to be fit for purpose and involves more than arbitrary choices.

The intention of this book is to illustrate the correct procedure for basic design of installations from initial assessment to final commissioning. It will also be of use to candidates studying for a C&G 2391-20 Design qualification.

This edition has been revised to serve as an accompaniment to the new City & Guilds scheme and has been brought fully up-to-date with the 17th Edition IEE Wiring Regulations.

Brian Scaddan, April 2011

CHAPTER 1

Design

Important terms/topics covered in this chapter:

- Assessment of general characteristics
- Basic protection
- Fault protection
- IP and IK codes
- The earth fault loop path
- Supplementary equipotential bonding
- Overcurrent
- Let-through-energy
- Discrimination
- Undervoltage
- Overvoltage
- Isolation and switching
- Design current
- Diversity
- Nominal rating of protection
- Rating factors
- Current carrying capacity of conductors
- Voltage drop
- Shock risk and thermal constraints.

At the end of this chapter the reader should,

- understand the need to assess all relevant characteristics of the installation,
- have reinforced their knowledge of Basic and Fault protection and how such protection is achieved,
- understand how 'let-through-energy' can cause damage to cables,
- be able to distinguish between 'Isolation' and 'Switching',
- be able to determine suitable wiring systems for particular applications,

IEE Wiring Regulations. DOI: 10.1016/B978-0-08-096914-5.10001-9

- be able to carry out basic design calculations to determine cable sizes,
- recognize various types of installation diagram.

Any design to the 17th Edition of the IEE Wiring Regulations BS 7671 must be primarily concerned with the safety of persons, property and livestock. All other considerations such as operation, maintenance, aesthetics, etc., while forming an essential part of the design, should never compromise the safety of the installation.

The selection of appropriate systems and associated equipment and accessories is an integral part of the design procedure, and as such cannot be addressed in isolation. For example, the choice of a particular type of protective device may have a considerable effect on the calculation of cable size or shock risk, or the integrity of conductor insulation under fault conditions.

Perhaps the most difficult installations to design are those involving additions and/or alterations to existing systems, especially where no original details are available, and those where there is a change of usage or a refurbishment of a premises, together with a requirement to utilize as much of the existing wiring system as possible.

So, let us investigate those parts of the Wiring Regulations that need to be considered in the early stages of the design procedure.

ASSESSMENT OF GENERAL CHARACTERISTICS

Regardless of whether the installation is a whole one, an addition, or an alteration, there will always be certain design criteria to be considered before calculations are carried out. Part 3 of the 17th Edition, 'Assessment of General Characteristics', indicates six main headings under which these considerations should be addressed. These are:

1. Purpose, supplies and structure
2. External influences
3. Compatibility

4. Maintainability
5. Recognized safety services
6. Assessment of continuity of service.

Let us look at these headings in a little more detail.

Purpose, supplies and structure

- For a new design, will the installation be suitable for its intended purpose?
- For a change of usage, is the installation being used for its intended purpose?
- If not, can it be used safely and effectively for any other purpose?
- Has the maximum demand been evaluated?
- Can diversity be taken into account?
- Are the supply and earthing characteristics suitable?
- Are the methods for protection for safety appropriate?
- If standby or safety supplies are used, are they reliable?
- Are the installation circuits arranged to avoid danger and facilitate safe operation?

External influences

Appendix 5 of the IEE Regulations classifies external influences which may affect an installation. This classification is divided into three sections, the environment (A), how that environment is utilized (B) and construction of buildings (C). The nature of any influence within each section is also represented by a number. Table 1.1 gives examples of the classification.

With external influences included on drawings and in specifications, installations and materials used can be designed accordingly.

Table 1.1   Examples of Classifications of External Influences

Environment	Utilization	Building
Water	Capability	Materials
AD6 Waves	**BA3** Handicapped	**CA1** Non-combustible

Compatibility

It is of great importance to ensure that damage to, or mal-operation of, equipment cannot be caused by harmful effects generated by other equipment even under normal working conditions. For example, MIMS cable should not be used in conjunction with discharge lighting, as the insulation can break down when subjected to the high starting voltages; the operation of residual current devices (RCDs) may be impaired by the magnetic fields of other equipment; computers, PLCs, etc. may be affected by normal earth leakage currents from other circuits.

Maintainability

The Electricity at Work Regulations 1989 require every system to be maintained such as to prevent danger; consequently, all installations require maintaining, some more than others, and due account of the frequency and quality of maintenance must be taken at the design stage. It is usually the industrial installations that are mostly affected by the need for regular maintenance, and hence, consultation with those responsible for the work is essential in order to ensure that all testing, maintenance and repair can be effectively and safely carried out. The following example may serve to illustrate an approach to consideration of design criteria with regard to a change of usage.

Example 1.1

A vacant two-storey light industrial workshop, 12 years old, is to be taken over and used as a Scout/Guide HQ. New shower facilities are to be provided. The supply is three-phase 400/230 V and the earthing system is TN-S.

The existing electrical installation on both floors comprises steel trunking at a height of 2.5 m around all perimeter walls, with steel conduit, to all socket outlets and switches (metal-clad), to numerous isolators and switch-fuses once used to control single- and three-phase machinery, and to the lighting which comprises fluorescent luminaires suspended by chains from the ceilings. The ground floor is to be used as the main activity area and part of the top floor at one end is to be converted to house separate male and female toilet and shower facilities accommodating two 8 kW/230 V shower units in each area.

If the existing electrical installation has been tested and inspected and shown to be safe:

1. Outline the design criteria, having regard for the new usage, for
   (a) The existing wiring system, and
   (b) The wiring to the new showers.
2. What would be the total assumed current demand of the shower units?

Suggested approach/solution

1(a) Existing system

Purpose, supplies and structure. Clearly the purpose for which the installation was intended has changed; however, the new usage is unlikely, in all but a few instances, to have a detrimental effect on the existing system. It will certainly be under-loaded; nevertheless this does not preclude the need to assess the maximum demand.

The supply and earthing arrangements will be satisfactory, but there may be a need to alter the arrangement of the installation, in order to rebalance the load across the phases now that machinery is no longer present.

External influences. The new shower area will probably have a classification AD3 or 4 and will be subject to Section 701, IEE Regulations. Ideally all metal conduit and trunking should be removed together with any socket outlets within 3 m of the boundary of zone 1. The trunking could be replaced with PVC; alternatively it could be boxed in using insulating material and screw-on lids to enable access. It could be argued that no action is necessary as it is above 2.25 m and therefore outside of all the zones. Suspended fluorescent fittings should be replaced with the enclosed variety, with control switches preferably located outside the area.

The activities in the ground-floor area will almost certainly involve various ball games, giving it a classification of AG2 (medium impact). Conduit drops are probably suitable, but old isolators and switch-fuses should be removed, and luminaires fixed to the ceiling and caged, or be replaced with suitably caged spotlights on side walls at high level.

As the whole building utilization can now be classified as BA2 (children), it is probably wise to provide additional protection against shock by installing 30 mA RCDs on all circuits.

Compatibility. Unlikely to be any compatibility problems with the new usage.

Maintainability. Mainly periodic test and inspection with some maintenance of lighting, hence suitable access equipment should be available, together with spare lamps and tubes. Lamp disposal facilities should be considered. A maintenance programme should be in place and all safety and protective measures should be effective throughout the intended life of the installation.

1(b) New shower area (BS 7671 Section 701)

Purpose, supplies and structure. As this is a new addition, the installation will be designed to fulfil all the requirements for which it is intended. The supply and earthing system should be suitable, but a measurement of the prospective fault current (PFC) and Z_e should be taken. The loading of the showers will have been accounted for during the assessment of maximum demand.

In the unlikely event of original design and installation details being available, it may be possible to utilize the existing trunking without exceeding space factors or de-rating cables due to the application of grouping factors. However, it is more probable that a re-evaluation of the trunking installation would need to be undertaken, or alternatively, the installation of a completely separate system. Whichever the method adopted, a distribution circuit supplying a four-way distribution board located outside the area would be appropriate, the final circuits to each shower being run via individual control switches also outside, and thence to the units using a PVC conduit system. Protection against shock would be by basic protection (insulation and barriers and enclosures) and fault protection (protective earthing, protective equipotential bonding and automatic disconnection); additional protection would be provided by RCDs/RCBOs.

External influences. These have already been addressed in above.

Compatibility. There will be no incompatibility between any equipment in this area.

Maintainability. Afforded by the individual switches and/or circuit breakers allowing isolation to maintain or repair/replace defective units.

2 Total assumed current demand

Design current I_b for each unit $= 8000/230 = 35$ A applying diversity:

1st unit	100% of 35=35
2nd unit	100% of 35=35
3rd unit	25% of 35=8.75
4th unit	25% of 35=8.75
Total assumed current demand=87.5 A	

As an answer to a C&G 2400 examination question, this suggested approach is more comprehensive than time constraints would allow, and hence an abbreviated form is acceptable. The solutions to the questions for Chapter 3 of this book illustrate such shortened answers.

PROTECTION FOR SAFETY

Part 4 of the 17th Edition details the methods and applications of *protection for safety,* and consideration of these details must be made as part of the design procedure. Areas that the designer needs to address are: protection against shock, thermal effects, overcurrent, overload, fault current, undervoltage, overvoltage, and the requirements for isolation and switching. Let us now deal, in broad terms, with each of these areas.

PROTECTION AGAINST ELECTRIC SHOCK

There are two ways that persons or livestock may be exposed to the effects of electric shock; these are (a) by touching live parts of electrical equipment, or (b) by touching exposed-conductive parts of electrical equipment or systems, which have been made live by a fault. Table 1.2 indicates the common methods of protecting against either of these situations.

Insulation or barriers and enclosures (Basic protection)

One method used to protect against contact with live parts is to insulate or house them in enclosures and/or place them behind barriers. In order to ensure that such protection will be satisfactory, the enclosures/barriers must conform to BS EN 60529, commonly referred to as the Index of Protection (IP) code. This details the amount of protection an enclosure can offer to the ingress of mechanical objects, foreign solid bodies and moisture. Table 1.3 (see page 10) shows part of the IP code. The X in a code simply means that protection is not specified; for example, in the code IP2X, only the protection against mechanical objects is specified, not moisture. Also, protection for wiring systems against external mechanical impact needs to be considered. Reference should be made to BS EN 62262, the IK code (Table 1.4, see page 11).

Protective earthing, protective equipotential bonding and automatic disconnection in case of a fault (Fault protection)

As Table 1.2 indicates, this method is the most common method of providing Fault protection, and hence it is important to expand on this topic.

There are two basic ways of receiving an electric shock by contact with conductive parts made live due to a fault:

1. Via parts of the body and the general mass of earth (typically hands and feet) or
2. Via parts of the body and simultaneously accessible *exposed and extraneous conductive parts* (typically hand to hand) – see Figure 1.1.

Clearly, the conditions shown in Figure 1.1 would provide no protection, as the installation is not earthed. However, if it can be ensured that protective devices operate fast enough by providing low impedance paths for earth fault currents, and that main protective bonding is carried out, then the magnitude and duration of earth faults will be reduced to such a level as not to cause danger.

Table 1.2 Common Methods of Protection Against Shock

Protection by	Protective Method	Applications and Comments
SELV (separated extra low voltage)	Basic and fault protection	Used for circuits in environments such as bathrooms, swimming pools, restrictive conductive locations, agricultural and horticultural situations, and for 25 V hand lamps in damp situations on construction sites. Also useful for circuits in schools, or college laboratories.
Insulation of live parts	Basic protection	This is simply 'basic insulation'.
Barriers and enclosures	Basic protection	Except where otherwise specified, such as swimming pools, hot air saunas, etc., placing LIVE PARTS behind barriers or in enclosures to at least IP2X is the norm. Two exceptions to this are: 1. Accessible horizontal top surfaces of, for example, distribution boards or consumer units, where the protection must be to at least IP4X and 2. Where a larger opening than IP2X is necessary, for example entry to lampholders where replacement of lamps is needed. Access past a barrier or into an enclosure should only be possible by the use of a tool, or after the supply has been disconnected, or if there is an intermediate barrier to at least IP2X. This does not apply to ceiling roses or ceiling switches with screw-on lids.
Obstacles	Basic protection	Restricted to areas only accessible to skilled persons, for example sub-stations with open fronted busbar chambers, etc.
Placing out of reach	Basic protection	Restricted to areas only accessible to skilled persons, e.g. sub-stations with open fronted busbar chambers, etc. Overhead travelling cranes or overhead lines.
RCDs (residual current devices)	Basic protection	These may only be used as additional protection, and must have an operating current of 30 mA or less, and an operating time of 40 ms or less at a residual current of $5 \times I_{\Delta n}$.
	Fault protection	Used where the loop impedance requirements cannot be met or for protecting socket outlet circuits supplying portable equipment used outdoors.
		Preferred method of earth fault protection for TT systems.

(Continued)

Table 1.2 Common Methods of Protection Against Shock—Cont'd

Protection by	Protective Method	Applications and Comments
Earthing, equipotential bonding and automatic disconnection of supply	Fault protection	The most common method in use. Relies on the co-ordination of the characteristics of the earthing, impedance of circuits, and operation of protective devices such that no danger is caused by earth faults occurring anywhere in the installation.
Class II equipment	Fault protection	Sometimes referred to as double insulated equipment and marked with the BS symbol ☐.
Non-conducting location	Fault protection	Rarely used – only for very special installations under strict supervision.
Earth-free local equipotential bonding	Fault protection	Rarely used – only for very special installations under strict supervision.
Electrical separation	Fault protection	Rarely used – only for very special installations under strict supervision. However, a domestic shaver point is an example of this method for one item of equipment.

Table 1.3 IP Codes

First Numeral:	Mechanical Protection
0	No protection of persons against contact with live or moving parts inside the enclosure. No protection of equipment against ingress of solid foreign bodies.
1	Protection against accidental or inadvertent contact with live or moving parts inside the enclosure by a large surface of the human body, for example, a hand, not for protection against deliberate access to such parts. Protection against ingress of large solid foreign bodies.
2	Protection against contact with live or moving parts inside the enclosure by fingers. Protection against ingress of medium-sized solid foreign bodies.
3	Protection against contact with live or moving parts inside the enclosure by tools, wires or such objects of thickness greater than 2.5 mm. Protection against ingress of small foreign bodies.
4	Protection against contact with live or moving parts inside the enclosure by tools, wires or such objects of thickness greater than 1 mm. Protection against ingress of small foreign bodies.

Table 1.3 IP Codes—Cont'd

First Numeral:	Mechanical Protection
5	Complete protection against contact with live or moving parts inside the enclosures. Protection against harmful deposits of dust. The ingress of dust is not totally prevented, but dust cannot enter in an amount sufficient to interfere with satisfactory operation of the equipment enclosed.
6	Complete protection against contact with live or moving parts inside the enclosures. Protection against ingress of dust.
Second Numeral:	Liquid Protection
0	No protection.
1	Protection against drops of condensed water. Drops of condensed water falling on the enclosure shall have no effect.
2	Protection against drops of liquid. Drops of falling liquid shall have no harmful effect when the enclosure is tilted at any angle up to 15° from the vertical.
3	Protection against rain. Water falling in rain at an angle equal to or smaller than 60° with respect to the vertical shall have no harmful effect.
4	Protection against splashing. Liquid splashed from any direction shall have no harmful effect.
5	Protection against water jets. Water projected by a nozzle from any direction under stated conditions shall have no harmful effect.
6	Protection against conditions on ships' decks (deck with watertight equipment). Water from heavy seas shall not enter the enclosures under prescribed conditions.
7	Protection against immersion in water. It must not be possible for water to enter the enclosure under stated conditions of pressure and time.
8	Protection against indefinite immersion in water under specified pressure. It must not be possible for water to enter the enclosure.

Table 1.4 IK Codes Protection Against Mechanical Impact

Code		
00		No protection
01 to 05		Impact 1 joule

(Continued)

Table 1.4	IK Codes Protection Against Mechanical Impact—Cont'd	
Code		
06	500 g / 20 cm	Impact 1 joule
07	500 g / 40 cm	Impact 2 joules
08	1.7 kg / 29.5 cm	Impact 5 joules
09	5 kg / 20 cm	Impact 10 joules
10	5 kg / 40 cm	Impact 20 joules

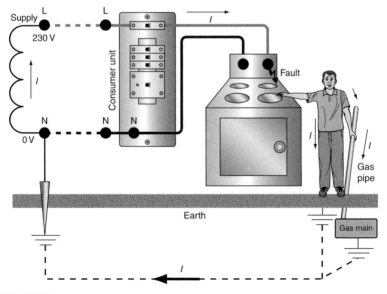

FIGURE 1.1 Shock path.

The disconnection times for final circuits not exceeding 32 A is 0.4 s and for distribution circuits and final circuits over 32 A is 5 s. For TT systems these times are 0.2 s and 1 s.

The connection of protective bonding conductors has the effect of creating a zone in which, under earth fault conditions, all exposed and extraneous conductive parts rise to a substantially equal potential. There may be differences in potential between simultaneously accessible conductive parts, but provided the design and installation are correct, the level of shock voltage will not be harmful.

Figure 1.2 shows the earth fault system which provides Fault protection.

The low impedance path for fault currents, the *earth fault loop path*, comprises that part of the system external to the installation, i.e. the impedance of the supply transformer, distributor and service cables Z_e, and the resistance of the line conductor R_1 and circuit protective conductor (cpc) R_2, of the circuit concerned.

The total value of loop impedance Z_s is therefore the sum of these values:

$$Z_s = Z_e + (R_1 + R_2)\,\Omega$$

Provided that this value of Z_s does not exceed the maximum value given for the protective device in question in Tables 41.2, 41.3 or 41.4 of the Regulations, the protection will operate within the prescribed time limits.

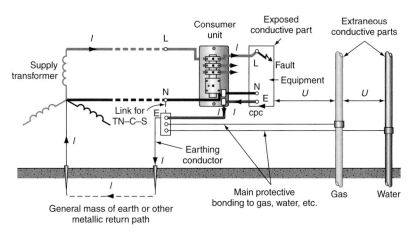

FIGURE 1.2   Earth fault loop path.

It must be noted that the actual value of $(R_1 + R_2)$ is determined from:

$$\frac{\text{Tabulated value of} \left(R_1 + R_2 \right) \times \text{Circuit length} \times \text{Multiplier}}{1000}$$

Note

The multiplier corrects the resistance at 20c to the value at conductor operating temperature.

External loop impedance Z_e

The designer obviously has some measure of control over the values of R_1 and R_2, but the value of Z_e can present a problem when the premises, and hence the installation within it, are at drawing-board stage. Clearly Z_e cannot be measured, and although a test made in an adjacent installation would give some indication of a likely value, the only recourse would either be to request supply network details from the Distribution Network Operator (DNO) and calculate the value of Z_e, or use the maximum likely values quoted by the DNOs, which are:

TT system	$21\,\Omega$
TN-S system	$0.8\,\Omega$
TN-C-S system	$0.35\,\Omega$

These values are pessimistically high and may cause difficulty in even beginning a design calculation. For example, calculating the size of conductors (considering shock risk) for, say, a distribution circuit cable protected by a 160 A, BS 88 fuse and supplied via a TNC-S system, would present great difficulties, as the maximum value of Z_s (Table 41.4(a)) for such a fuse is $0.25\,\Omega$ and the quoted likely value of Z_e is $0.35\,\Omega$. In this case the DNO would need to be consulted.

Supplementary equipotential bonding

This still remains a contentious issue even though the Regulations are quite clear on the matter. Supplementary bonding is used as Additional protection to Fault protection and required under the following conditions:

1. When the requirements for loop impedance and associated discon-nection times cannot be met (RCDs may be installed as an alterna-tive), and
2. The location is an area of increased risk such as detailed in Part 7 of the Regulations, e.g. bathrooms, etc. and swimming pools (see also Chapter 3).

PROTECTION AGAINST THERMAL EFFECTS (IEE REGULATIONS CHAPTER 42)

The provision of such protection requires, in the main, a commonsense approach. Basically, ensure that electrical equipment that generates heat is so placed as to avoid harmful effects on surrounding combustible mate-rial. Terminate or join all live conductors in approved enclosures, and where electrical equipment contains in excess of 25 litres of flammable liquid, make provision to prevent the spread of such liquid, for example a retaining wall round an oil-filled transformer.

In order to protect against burns from equipment not subject to a Har-monized Document limiting temperature, the designer should conform to the requirements of Table 42.1, IEE Regulations.

Section 422 of this chapter deals with locations and situations where there may be a particular risk of fire. These would include locations where combustible materials are stored or could collect and where a risk of ignition exists. This chapter does not include locations where there is a risk of explosion.

PROTECTION AGAINST OVERCURRENT

The term overcurrent may be sub-divided into:

1. Overload current and
2. Fault current.

The latter is further sub-divided into:

(a) Short-circuit current (between live conductors) and
(b) Earth fault current (between line and earth).

Overloads are overcurrents occurring in healthy circuits and caused by, for example, motor starting, inrush currents, motor stalling, connection of more loads to a circuit than it is designed for, etc.

Fault currents, on the other hand, typically occur when there is mechanical damage to circuits and/or accessories causing insulation failure or breakdown leading to 'bridging' of conductors. The impedance of such a 'bridge' is assumed to be negligible.

Clearly, significant overcurrents should not be allowed to persist for any length of time, as damage will occur to conductors and insulation.

Table 1.5 indicates some of the common types of protective device used to protect electrical equipment during the presence of overcurrents and fault currents.

PROTECTION AGAINST OVERLOAD

Protective devices used for this purpose have to be selected to conform with the following requirements:

1. The nominal setting of the device I_n must be greater than or equal to the design current I_b:

$$I_n \geq I_b$$

2. The current-carrying capacity of the conductors I_z must be greater than or equal to the nominal setting of the device I_n:

$$I_z \geq I_n$$

3. The current causing operation of the device I_2 must be less than or equal to 1.45 times the current-carrying capacity of the conductors I_z:

$$I_2 \leq 1.45 \times I_z$$

For fuses to BS 88 and BS 1361, and MCBs or CBs, compliance with (2) above automatically gives compliance with (3). For fuses to BS 3036

Table 1.5   Commonly Used Protective Devices

Device	Application	Comments
Semi-enclosed re-wireable fuse BS 3036	Mainly domestic consumer units	Gradually being replaced by other types of protection. Its high fusing factor results in lower cable current carrying capacity or, conversely, larger cable sizes.
		Does not offer good short-circuit current protection.
		Ranges from 5 A to 200 A.
HBC fuse links BS 88-6 and BS EN 60269-2	Mainly commercial and industrial use	Give excellent short-circuit current protection. Does not cause cable de-rating. 'M' types used for motor protection. Ranges from 2 A to 1200 A.
HBC fuse links BS 1361	House service and consumer unit fuses	Not popular for use in consumer units; however, gives good short-circuit current protection, and does not result in cable de-rating.
		Ranges from 5 A to 100 A.
MCBs and CBs (miniature circuit breakers) BS 3871, now superseded by BS EN 60898 CBs	Domestic consumer units and commercial/industrial distribution boards	Very popular due to ease of operation. Some varieties have locking-off facilities. Range from 1 A to 63 A single and three phase. Old types 1, 2, 3 and 4 now replaced by types B, C and D with breaking capacities from 3 kA to 25 kA.
MCCBs (moulded case circuit breakers) BS EN 60947-2	Industrial situations where high current and breaking capacities are required	Breaking capacity, 22–50 kA in ranges 16–1200 A. 2, 3 and 4 pole types available.

(re-wireable) compliance with (3) is achieved if the nominal setting of the device I_n is less than or equal to $0.725 \times I_z$:

$$I_n \leq 0.725 \times I_z$$

This is due to the fact that a re-wireable fuse has a fusing factor of 2, and $1.45/2 = 0.725$.

Overload devices should be located at points in a circuit where there is a reduction in conductor size or anywhere along the length of a conductor, providing there are no branch circuits. The Regulations indicate circumstances under which overload protection may be omitted; one such example is when the characteristics of the load are not likely to cause an overload, hence there is no need to provide protection at a ceiling rose for the pendant drop.

PROTECTION AGAINST FAULT CURRENT

Short-circuit current

When a 'bridge' of negligible impedance occurs between live conductors (remember, a neutral conductor is a live conductor) the short-circuit current that could flow is known as the 'prospective short-circuit current' (PSCC), and any device installed to protect against such a current must be able to break and in the case of a circuit breaker, make the PSCC at the point at which it is installed without the scattering of hot particles or damage to surrounding materials and equipment. It is clearly important therefore to select protective devices that can meet this requirement.

It is perhaps wise to look in a little more detail at this topic. Figure 1.3 shows PSCC over one half-cycle; t_1 is the time taken to reach 'cut-off' when the current is interrupted, and t_2 the total time taken from start of fault to extinguishing of the arc.

During the 'pre-arcing' time t_1, electrical energy of considerable proportions is passing through the protective device into the conductors. This is known as the 'pre-arcing let-through' energy and is given by $(I_f)^2 t_1$ where I_f is the short-circuit current at 'cut-off'.

The total amount of energy let-through into the conductors is given by $(I_f)^2 t_2$ in Figure 1.4.

For faults up to 5 s duration, the amount of heat and mechanical energy that a conductor can withstand is given by $k^2 s^2$, where k is a factor dependent on the conductor and insulation materials (tabulated in the Regulations), and s is the conductor csa. Provided the energy let-through by the protective device does not exceed the energy withstand of the conductor, no damage will occur. Hence, the limiting situation is when $(I_f)^2 t = k^2 s^2$. If we now transpose this formula for t, we get $t = k^2 s^2/(I_f)^2$, which is the maximum disconnection time (t in seconds).

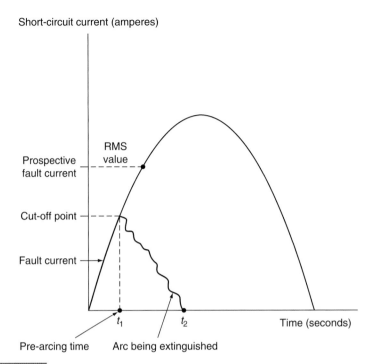

FIGURE 1.3 Pre-arcing let-through.

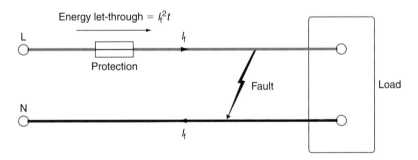

FIGURE 1.4 Pre-arcing let-through.

When an installation is being designed, the PSCC at each relevant point in the installation has to be determined, unless the breaking capacity of the lowest rated fuse in the system is greater than the PSCC at the intake position. For supplies up to 100 A the supply authorities quote a value of PSCC, at the point at which the service cable is joined to the distributor cable, of 16 kA. This value will decrease significantly over only a short length of service cable.

Earth fault current

We have already discussed this topic with regard to shock risk, and although the protective device may operate fast enough to prevent shock, it has to be ascertained that the duration of the fault, however small, is such that no damage to conductors or insulation will result. This may be verified in two ways:

1. If the protective conductor conforms to the requirements of Table 54.7 (IEE Regulations), or if
2. The csa of the protective conductor is not less than that calculated by use of the formula:

$$s = \frac{\sqrt{I^2 t}}{k}$$

which is another rearrangement of $I^2 t = k^2 s^2$.

For flat, twin and three-core cables the formula method of verification will be necessary, as the cpc incorporated in such cables is always smaller than the associated line conductor. It is often desirable when choosing a cpc size to use the calculation, as invariably the result leads to smaller cpcs and hence greater economy. This topic will be expanded further in the section 'Design Calculations'.

Discrimination

It is clearly important that, in the event of an overcurrent, the protection associated with the *circuit in question* should operate, and not other devices upstream. It is not enough to simply assume that a device one size lower will automatically discriminate with one a size higher. All depends on the 'let-through' energy of the devices. If the *total* 'let-through' energy of the lower rated device does not exceed the *pre-arcing* 'let-through' energy of the higher rated device, then discrimination is achieved. Table 1.6 shows the 'let-through' values for a range of BS 88 fuse links, and illustrates the fact that devices of consecutive ratings do not necessarily discriminate. For example, a 6 A fuse will *not* discriminate with a 10 A fuse.

Table 1.6 $(I_t)^2{}_t$ Characteristics: 2–800 A Fuse Links. Discrimination is Achieved If the Total $(I_t)^2{}_t$ of the Minor Fuse Does Not Exceed the Pre-arcing $(I_t)^2{}_t$ of the Major Fuse

Rating (A)	$I_t^2 t$ Pre-arcing	$I_t^2 t$ Total at 415 V
2	0.9	17
4	4	12
6	16	59
10	56	170
16	190	580
20	310	810
25	630	1700
32	1200	2800
40	2000	6000
50	3600	11 000
63	6500	14 000
80	13 000	36 000
100	24 000	66 000
125	34 000	120 000
160	80 000	260 000
200	140 000	400 000
250	230 000	560 000
315	360 000	920 000
350	550 000	1 300 000
400	800 000	2 300 000
450	700 000	1 400 000
500	900 000	1 800 000
630	2 200 000	4 500 000
700	2 500 000	5 000 000
800	4 300 000	10 000 000

PROTECTION AGAINST UNDERVOLTAGE (IEE REGULATIONS SECTION 445)

In the event of a loss of or significant drop in voltage, protection should be available to prevent either damage or danger when the supply is restored. This situation is most commonly encountered in motor circuits, and in this case the protection is provided by the contactor coil via the control circuit. If there is likely to be damage or danger due to undervoltage, standby supplies could be installed and, in the case of computer systems, uninterruptible power supplies (UPS). Switching on of very large loads can have the effect of causing such undervoltages.

PROTECTION AGAINST OVERVOLTAGE (IEE REGULATIONS SECTIONS 442 AND 443)

This chapter deals with the requirements of an electrical installation to withstand overvoltages caused by: (1) transient overvoltages of atmospheric origin, and (2) switching surges within the installation. It is unlikely that installations in the UK will be affected by the requirements of item (1) as the number of thunderstorm days per year is not likely to exceed 25. In the case of item (2), when highly inductive loads are switched, the sudden movement of associated magnetic fields can cause transient overvoltages.

ISOLATION AND SWITCHING

Let us first be clear about the difference between isolators and switches. An isolator is, by definition, 'A mechanical switching device which provides the function of cutting off, for reasons of safety, the supply to all or parts of an installation, from every source. A switch is a mechanical switching device which is capable of making, carrying and breaking normal load current, and some overcurrents. It may not break short-circuit currents'.

So, an isolator may be used for functional switching, but not usually vice versa. Basically an isolator is operated after all loads are switched off, in order to prevent energization while work is being carried out. Isolators are off-load devices, switches are on-load devices.

The IEE Regulations (Section 537) deal with this topic and in particular Isolation, Switching off for mechanical maintenance, Emergency switching, and Functional switching.

Tables 1.7–1.9 indicate some of the common devices and their uses.

Table 1.7 Common Types of Isolators and Switches

Device	Application	Comments
Isolator or disconnector	Performs the function of isolation	Not designed to be operated on load. Isolation can also be achieved by the removal of fuses, pulling plugs, etc.
Functional switch	Any situation where a load needs to be frequently operated, i.e. light switches, switches on socket outlets, etc.	A functional switch could be used as a means of isolation, i.e. a one-way light switch provides isolation for lamp replacement provided the switch is under the control of the person changing the lamp.
Switch-fuse	At the origin of an installation or controlling sub-mains or final circuits	These can perform the function of isolation while housing the circuit protective devices.
Fuse-switch	As for switch-fuse	Mainly used for higher current ratings and have their fuses as part of the moving switch blades.
Switch disconnector	Main switch on consumer units and distribution fuse boards	These are ON LOAD devices but can still perform the function of isolation.

Table 1.8 Common devices

Device	Isolation	Emergency	Function
Circuit breakers	Yes	Yes	Yes
RCDs	Yes	Yes	Yes
Isolating switches	Yes	Yes	Yes
Plugs and socket outlets	Yes	No	Yes
Ditto but over 32 A	Yes	No	No
Switched fused connection unit	Yes	Yes	Yes
Unswitched fused connection unit	Yes	No	No
Plug fuses	Yes	No	No
Cooker units	Yes	Yes	Yes

Table 1.9 Common Wiring Systems and Cable Types

System/Cable Type	Applications	Comments
1 Flat twin and three-core cable with cpc; PVC sheathed, PVC insulated, copper conductors	Domestic and commercial fixed wiring	Used clipped direct to surface or buried in plaster either directly or encased in oval conduit or top-hat section; also used in conjunction with PVC mini-trunking.
2 PVC mini-trunking	Domestic and commercial fixed wiring	Used with (1) above for neatness when surface wiring is required.
3 PVC conduit with single-core PVC insulated copper conductors	Commercial and light industrial	Easy to install, high impact, vermin proof, self-extinguishing, good in corrosive situations. When used with 'all insulated' accessories provides a degree of Fault protection on the system.
4 PVC trunking: square, rectangular, skirting, dado, cornice, angled bench. With single-core PVC insulated copper conductors	Domestic, commercial and light industrial	When used with all insulated accessories provides a degree of Fault protection on the system. Some forms come pre-wired with copper busbars and socket outlets. Segregated compartment type good for housing different band circuits.
5 Steel conduit and trunking with single-core PVC insulated copper conductors	Light and heavy industry, areas subject to vandalism	Black enamelled conduit and painted trunking used in non-corrosive, dry environments. Galvanized finish good for moist/damp or corrosive situations. May be used as cpc, though separate one is preferred.
6 Busbar trunking	Light and heavy industry, rising mains in tall buildings	Overhead plug-in type ideal for areas where machinery may need to be moved. Arranged in a ring system with section switches, provides flexibility where regular maintenance is required.
7 Mineral insulated copper sheathed (MICS) cable exposed to touch or PVC covered. Clipped direct to a surface or perforated tray or in trunking or ducts	All industrial areas, especially chemical works, boiler houses, petrol filling stations, etc.; where harsh conditions exist such as extremes of heat, moisture, corrosion, etc., also used for fire alarm circuits	Very durable, long-lasting, can take considerable impact before failing. Conductor current-carrying capacity greater than same in other cables. May be run with circuits of different categories in unsegregated trunking. Cable reference system as follows:

Table 1.9   Common Wiring Systems and Cable Types—Cont'd

System/Cable Type	Applications	Comments
		CC – bare copper sheathed MI cable
		V – PVC covered
		M – low smoke and fume (LSF) material covered
		L – light duty (500V)
		H – heavy duty (750V)
		Hence a two-core 2.5mm² light duty MI cable with PVC oversheath would be shown: CCV 2L 2.5.
8 F.P. 200. PVC sheathed aluminium screened silicon rubber insulated, copper conductors. Clipped direct to surface or on perforated tray or run in trunking or ducts	Fire alarm and emergency lighting circuits	Specially designed to withstand fire. May be run with circuits of different categories in non-segregated trunking.
9 Steel wire armoured. PVC insulated, PVC sheathed with copper conductors, clipped direct to a surface or on cable tray or in ducts or underground	Industrial areas, construction sites, underground supplies, etc.	Combines a certain amount of flexibility with mechanical strength and durability.
10 As above but insulation is XLPE. Cross (X) linked (L) poly (P) ethylene (E)	For use in high temperature areas	As above.
11 HOFR sheathed cables (heat, oil, flame retardant)	All areas where there is a risk of damage by heat, oil or flame	These are usually flexible cords.

DESIGN CALCULATIONS

Basically, all designs follow the same procedure:

1. Assessment of general characteristics
2. Determination of design current I_b
3. Selection of protective device having nominal rating or setting I_n
4. Selection of appropriate rating factors
5. Calculation of tabulated conductor current I_t
6. Selection of suitable conductor size
7. Calculation of voltage drop
8. Evaluation of shock risk
9. Evaluation of thermal risks to conductors.

Let us now consider these steps in greater detail. We have already dealt with 'assessment of general characteristics', and clearly one result of such assessment will be the determination of the type and disposition of the installation circuits. Table 1.9 gives details of commonly installed wiring systems and cable types. Having made the choice of system and cable type, the next stage is to determine the design current.

Design current I_b

This is defined as '*the magnitude of the current to be carried by a circuit in normal service*', and is either determined directly from manufacturers' details or calculated using the following formulae:

Single phase:

$$I_b = \frac{P}{V} \quad \text{or} \quad \frac{P}{V \times \text{Eff}\% \times \text{PF}}$$

Three phase:

$$I_b = \frac{P}{\sqrt{3} \times V_L} \text{or} \frac{P}{\sqrt{3} \times V_L \times \text{Eff}\% \times \text{PF}}$$

where:

P = power in watts

V = line to neutral voltage in volts

V_L = line to line voltage in volts

Eff% = efficiency

PF = power factor.

Diversity

The application of diversity to an installation permits, by assuming that not all loads will be energized at the same time, a reduction in main or distribution circuit cable sizes. The IEE Regulations Guidance Notes or On-Site Guide tabulate diversity in the form of percentages of full load for various circuits in a range of installations. However, it is for the designer to make a careful judgement as to the exact level of diversity to be applied.

Nominal rating or setting of protection I_n

We have seen earlier that the first requirement for I_n is that it should be greater than or equal to I_b. We can select for this condition from IEE Regulations Tables 41.2, 41.3 or 41.4. For types and sizes outside the scope of these tables, details from the manufacturer will need to be sought.

Rating factors

There are several conditions which may have an adverse effect on conductors and insulation, and in order to protect against this, rating factors (CFs) are applied. These are:

C_a	Factor for ambient air and ground temperature	(From IEE Regulations Tables 4B1, 4B2 or 4B3)
C_g	Factor for groups of cables	(From IEE Regulations Table 4C1 to 4C5)
C_f	Factor: if BS 3036 re-wireable fuse is used	(Factor is 0.725)
C_i	Factor if cable is surrounded by thermally insulating material	(IEE Regulations, Table 52.2)

Application of rating factors

The factors are applied as divisors to the setting of the protection I_n; the resulting value should be less than or equal to the tabulated current-carrying capacity I_t of the conductor to be chosen.

It is unlikely that all of the adverse conditions would prevail at the same time along the whole length of the cable run and hence only the relevant factors would be applied. A blanket application of correction factors can result in unrealistically large conductor sizes, so consider the following:

1. If the cable in Figure 1.5 ran for the whole of its length, grouped with others of the same size in a high ambient temperature, and was totally surrounded with thermal insulation, it would seem logical to apply all the CFs, as they all affect the whole cable run. Certainly the factors for the BS 3036 fuse, grouping and thermal insulation should be used. However, it is doubtful if the ambient temperature will have any effect on the cable, as the thermal insulation, if it is efficient, will prevent heat reaching the cable. Hence apply C_g, C_f and C_i.

2. In Figure 1.6(a) the cable first runs grouped, then leaves the group and runs in high ambient temperature, and finally is enclosed in thermal insulation. We therefore have three different conditions, each affecting the cable in different areas. The BS 3036 fuse affects the whole cable run and therefore C_f must be used, but there is no need to apply all of the remaining factors as the worst one will automatically compensate for the others. The relevant

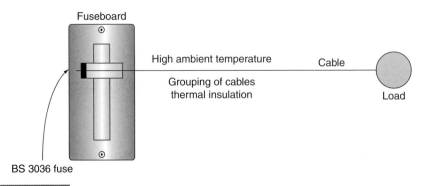

FIGURE 1.5   Conditions along cable route.

factors are shown in Figure 1.6(b) and apply only if $C_f = 0.725$ and $C_i = 0.5$. If protection was *not* by BS 3036 fuse, then apply only $C_i = 0.5$.

3. In Figure 1.7 a combination of cases 1 and 2 is considered. The effect of grouping and ambient temperature is $0.7 \times 0.97 = 0.679$. The factor for thermal insulation is still worse than this combination, and therefore C_i is the only one to be used.

Tabulated conductor current-carrying capacity I_t

$$I_n \geq \frac{I_n}{C_a \times C_g \times C_f \times C_i}$$

Remember, only the relevant factors are to be used!

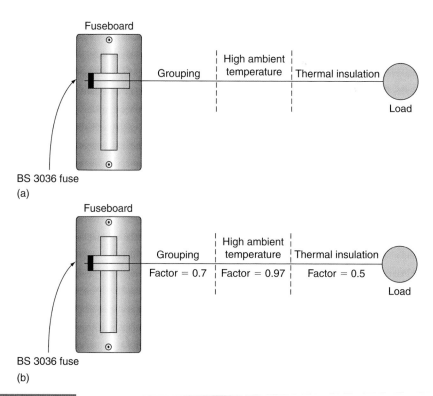

FIGURE 1.6   Conditions along cable route.

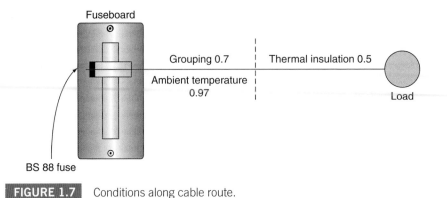

FIGURE 1.7   Conditions along cable route.

As we have seen when discussing overload protection, the IEE Regulations permit the omission of such protection in certain circumstances (433.3.1); in these circumstances, I_n is replaced by I_b and the formula becomes:

$$I_n \geq \frac{I_b}{C_a \times C_g \times C_f \times C_i}$$

Selection of suitable conductor size

During the early stages of the design, the external influences will have been considered, and a method of circuit installation chosen. Appendix 4, IEE Regulations Table 4A2 gives examples of installation methods, and it is important to select the appropriate method in the current rating tables. For example, from IEE Regulations Table 4D2A the tabulated current ratings I_t for reference method B are less than those for method C. Having selected the correct cable rating table and relevant reference method, the conductor size is determined to correspond with I_t.

Voltage drop

In many instances this may well be the most onerous condition to affect cable sizes. The Regulations require that the voltage at the terminals of fixed equipment should be greater than the lower limit permitted by the British Standard for that equipment, or in the absence of a British Standard, that the safe functioning of the equipment should not be impaired.

Table 1.10   Voltage Drop Values

	Lighting	Power
	3%	5%
230 V single phase	6.9 V	11.5 V
400 V three phase	12 V	20 V

These requirements are fulfilled if the voltage drop between the origin of the installation and any load point does not exceed the following values (IEE Regulations, Appendix 12) (Table 1.10).

Accompanying the cable current rating tables are tabulated values of voltage drop based on the milli-volts (mV) dropped for every ampere of design current (A), for every metre of conductor length (m), i.e.

Volt drop $= mV/A/m$

or fully translated with I_b for A and L (length in metres):

$$\text{Volt drop} = \frac{mV \times I_b \times \text{length}}{1000} \text{ volts}$$

For conductor sizes in excess of $16\,mm^2$ the impedance values of volt drop in the IEE Regulations tables, Appendix 4 (columns headed z) should be used. The columns headed r and x indicate the resistive and reactive components of the impedance values.

Evaluation of shock risk

This topic has been discussed earlier; suffice to say that the calculated value of loop impedance should not exceed the tabulated value quoted for the protective device in question.

Evaluation of thermal constraints

As we know, the 'let-through' energy of a protective device under fault conditions can be considerable and it is therefore necessary to ensure that the cpc is large enough, either by satisfying the requirements of

IEE Regulations Table 54.7 or by comparing its size with the minimum derived from the formula:

$$s = \frac{\sqrt{I^2 t}}{k}$$

where:

s = minimum csa of the cpc

I = fault current

t = disconnection time in seconds

k = factor taken from IEE Regulations Tables 54.2 to 54.6.

The following examples illustrate how this design procedure is put into practice.

Example 1.2

A consumer has asked to have a new 9 kW/230 V shower unit installed in a domestic premises. The existing eight-way consumer unit houses BS 3871 MCBs and supplies two ring final circuits, one cooker circuit, one immersion heater circuit and two lighting circuits, leaving two spare ways. The earthing system is TN–C–S with a measured value of Z_e of 0.18 Ω, and the length of the run from consumer unit to shower is approximately 28 m. The installation reference method is method C, and the ambient temperature will not exceed 30°C. If flat twin cable with cpc is to be used, calculate the minimum cable size.

Assessment of general characteristics

In this case, the major concern is the maximum demand. It will need to be ascertained whether or not the increased load can be accommodated by the consumer unit and the supplier's equipment.

Design current I_b (based on rated values)

$$I_b = \frac{P}{V} = \frac{9000}{230} = 39 \text{ A}$$

Choice and setting of protection

The type of MCB most commonly found in domestic installations over 10 years old is a BS 3871 Type 2, and the nearest European standard to this is a BS EN 60898 Type B. So from IEE Regulations Table 41.3, the protection would be a 40 A Type B CB with a corresponding maximum value of loop impedance Z_s of $1.15\,\Omega$.

Tabulated conductor current-carrying capacity I_t

As a shower is unlikely to cause an overload, I_b may be used instead of I_n:

$$I_t \geq \frac{I_b}{C_a \times C_g \times C_f \times C_i}$$

but as there are no rating factors,

$$I_t \geq I_b \quad \text{so} \quad I_t \geq 39\,\text{A}$$

Selection of conductor size

As the cable is to be PVC Twin with cpc, the conductor size will be selected from IEE Regulations Table 4D5 column 6. Hence I_t will be 47 A and the conductor size $6.0\,\text{mm}^2$.

Voltage drop

From IEE Regulations Table 4D5 column 7, the mV drop is 7.3, so:

$$\text{Volt drop} = \frac{\text{mV} \times I_b \times L}{1000} = \frac{7.3 \times 39 \times 28}{1000} = 7.97\,\text{V (acceptable)}$$

Whilst this may satisfy BS 7671, such a high value could cause inefficiency and the manufacturer should be consulted.

Evaluation for shock risk

The line conductor of the circuit has been calculated as $6.0\,\text{mm}^2$, and a twin cable of this size has a $2.5\,\text{mm}^2$ cpc. So, using the tabulated values of R_1 and R_2 given in the On-Site Guide, 28 m of cable would have a resistance under operating conditions of:

$$\frac{28 \times (3.08 + 7.41) \times 1.2}{1000} = 0.35\,\Omega$$

(1.2 = multiplier for 70°C conductor operating temperature) and as Z_e is 0.18, then:

$$Z_s = Z_e + R_1 + R_2 = 0.18 + 0.35 = 0.53\,\Omega$$

which is clearly less than the maximum value of 1.15.

Evaluation of thermal constraints

Fault current I is found from:

$$I = \frac{U_0}{Z_s}$$

where

U_0 = nominal line voltage to earth

Z_s = calculated value of loop impedance

$$I = \frac{230}{0.53} = 434\,\text{A}$$

t for 434 A from IEE Regulations curves, Figure 3.4 for a 40 A CB is less than 0.1 s. k from IEE Regulations Table 54.3 is 115.

$$s = \frac{\sqrt{I^2 t}}{k} = \frac{\sqrt{434^2 \times 0.1}}{115} = 1.2\,\text{mm}^2$$

which means that the 2.5 mm² cpc is perfectly adequate. It does not mean that a 1.2 mm² cpc could be used.

Hence, provided the extra demand can be accommodated, the new shower can be wired in 6.0 mm² flat twin cable with a 2.5 mm² cpc and protected by a 40 A BS EN 60898 Type B CB.

Example 1.3

Four industrial single-phase fan assisted process heaters are to be installed adjacent to each other in a factory. Each one is rated at 50 A/230 V. The furthest heater is some 32 m from a distribution board, housing BS 88 fuses, located at the intake position. It has been decided to supply the heaters with PVC singles in steel trunking (reference method B), and part of the run will be through an area where the ambient temperature may reach 35°C. The earthing system is TN–S with a measured Z_e of 0.3 Ω. There is spare capacity in the distribution board, and the maximum demand will not be exceeded. Calculate the minimum size of live conductors and cpc.

Calculations will be based on the furthest heater. Also, only one common cpc needs to be used (IEE Regulation 543.1.2).

Design current I_b

$$I_b = 50\,A$$

Type and setting of protection I_n

$I_n \geq I_b$ so, from IEE Regulations Table 41.4, a BS 88 50 A fuse would be used, with a corresponding maximum value of Z_s of 1.04 Ω (Figure 1.8).

Rating factors

As the circuits will be grouped and, for part of the length, run in a high ambient temperature, both C_a and C_g will need to be used.

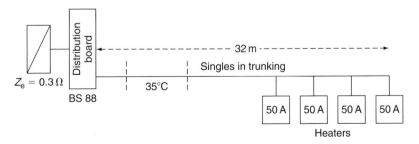

FIGURE 1.8   Diagram for example 1.3

C_a for 35°C	0.94 (Table 4B1)
C_g for four circuits	0.65 (Table 4C1)

Tabulated current-carrying capacity I_t

As the heaters are fan assisted, they are susceptible to overload, hence I_n is used:

$$I_t \geq \frac{I_n}{C_a \times C_g} \geq \frac{50}{0.94 \times 0.65} \geq 82 \,\text{A}$$

Selection of conductor size

From IEE Regulations Table 4D1A column 4, $I_t = 101 \,\text{A}$, and the conductor size is $25.0 \,\text{mm}^2$.

Voltage drop

From IEE Regulations Table 4D1B, the mV drop for $25.0 \,\text{mm}^2$ is $1.8 \,\text{mV}$.

$$\text{Volt drop} = \frac{1.8 \times 50 \times 32}{1000} = 2.88 \,\text{V} \,(\text{acceptable})$$

Evaluation of shock risk

In this case, as the conductors are singles, a cpc size has to be chosen either from IEE Regulations Table 54.7, or by calculation. The former

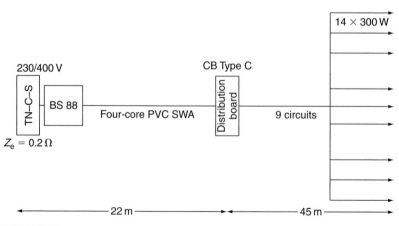

FIGURE 1.9 Diagram for example 1.4

method will produce a size of 16 mm², whereas calculation tends to produce considerably smaller sizes. The calculation involves the rearrangement of the formula:

$$Z_s = Z_e + \frac{(R_1 \times R_2) \times L \times 1.2}{1000}$$

to find the maximum value of R_2 and selecting a cpc size to suit. The value of Z_s used will be the tabulated maximum, which in this case is 1.04. The rearranged formula is:

$$R_2 = \frac{\left[(Z_s - Z_e) \times 1000\right]}{L \times 1.2} - R_1 = \frac{\left[(1.04 - 0.3) \times 1000\right]}{32 \times 1.2}$$
$$-0.727 \left(\text{from } R_1 + R_2 \text{ tables}\right) = 18.54\,\text{m'}$$

The nearest value to this maximum is 18.1 mΩ (from $R_1 + R_2$ tables) giving a cpc size of 1.0 mm². This will satisfy the shock risk requirements, but we will still have to know the actual value of Z_s, so:

$$Z_s = 0.3 + \frac{(0.727 + 18.1) \times 1.2 \times 32}{1000} = 1.0\,\Omega$$

Evaluation of thermal constraints

$$\text{Fault current } I = \frac{U_o}{Z_s} = \frac{230}{1} = 230\,\text{A}$$

t from 50 A BS 88 curve $= 3\,\text{s}$

$k = 115$ (IEE Regulations Table 54.3)

$$s = \frac{\sqrt{I^2 t}}{k} = \frac{\sqrt{230^2 \times 3}}{115} = 3.46\,\text{mm}^2$$

Hence, our 1.0 mm² cpc is too small to satisfy the thermal constraints, and hence a 4.0 mm² cpc would have to be used. So the heaters would be supplied using 25.0 mm² live conductors, a 4.0 mm² cpc and 50 A BS 88 protection.

Example 1.4

Part of the lighting installation in a new warehouse is to comprise a distribution circuit to a three-phase lighting distribution board from which nine single-phase final circuits are to be fed. The distribution circuit, protected by BS 88 fuses, is to be four-core PVC SWA cable and is 22 m long. The armouring will provide the function of the cpc. The distribution board will house BS EN 60898 Type C CBs, and each final circuit is to supply fourteen 300 W discharge luminaires. The longest run is 45 m, and the wiring system will be singles in trunking, the first few metres of which will house all nine final circuits. The earthing system is TN-C-S and the value of Z_s calculated to be 0.2 Ω. The ambient temperature will not exceed 30°C (see Fig. 1.9).

Determine all relevant cable/conductor sizes

Design current of each final circuit I_b

As each row comprises fourteen 300 W/230 V discharge fittings:

$$I_b = \frac{14 \times 300 \times 1.8}{230} = 32.8 \text{ A}$$

(the 1.8 is the multiplier for discharge lamps)

As the nine circuits will be balanced over three phases, each phase will feed three rows of fittings:

$$I_b \text{ per phase} = 3 \times 32.8 = 98.4 \text{ A}$$

Distribution circuit design current I_b

Distribution circuit I_b per phase = 98.4 A.

Nominal rating of protection I_n

$I_n \geq I_b$ so, from IEE Regulations Table 41.4, the protection will be 100 A with a maximum loop impedance Z_s of 0.42 Ω.

Rating factors

Not applicable.

Tabulated current-carrying capacity I_t

Discharge units do cause short duration overloads at start-up, so it is perhaps best to use I_n rather than I_b:

$$I_t \geq I_n \geq 100\,\mathrm{A}$$

Cable selection

From IEE Regulations Table 4D4A column 3, $I_t = 102\,\mathrm{A}$, giving a cable size of $25\,\mathrm{mm}^2$.

Voltage drop

From IEE Regulations Table 4D4B column 4, the mV drop is 1.5.

$$\text{Volt drop} = \frac{1.5 \times 98.4 \times 22}{1000} = 3.23\,\mathrm{V}\,(\text{acceptable})$$

This is the three-phase drop, the single phase being:

$$\frac{3.23}{\sqrt{3}} = 1.87\,\mathrm{V}$$

Evaluation of shock risk

Cable manufacturer's information shows that the resistance of the armouring on a $25\,\mathrm{mm}^2$ four-core cable is $2.1\,\mathrm{m}\Omega/\mathrm{m}$. Hence,

$$R_1 = 0.727 \ \mathrm{m}\Omega/\mathrm{m} \text{ and } R_2 = 2.1 \ \mathrm{m}\Omega/\mathrm{m}$$

$$Z_s = 0.2 + \frac{(0.727 + 2.1) \times 22 \times 1.2}{1000} = 0.274 \ \Omega$$

Clearly ok, as Z_s maximum is $0.42\,\Omega$.

Thermal constraints

$$I = \frac{U_o}{Z_s} = \frac{230}{0.274} = 839\,\mathrm{A}$$

$t = 0.7$ from BS 88 (curve for $100\,\mathrm{A}$)

$k = 51$ (IEE Regulations Table 54.4)

$$s = \frac{\sqrt{839^2 \times 0.7}}{51} = 13.76\,\text{mm}^2$$

Manufacturer's information gives the gross csa of 25 mm² four-core SWA cable as 76 mm². Hence the armouring provides a good cpc.

If we had chosen to use IEE Regulations Table 54.7 to determine the minimum size it would have resulted in:

$$s = \frac{16 \times k_1}{k_2} = \frac{16 \times 115}{51} = 36\,\text{mm}^2$$

which still results in a smaller size than will exist.

Final circuits design current I_b

$I_b = 32.8\,\text{A}$ (calculated previously).

Setting of protection I_n

From IEE Regulations Table 41.3, $I_n \geq I_b = 40\,\text{A}$ with a corresponding maximum value for Z_s of $0.57\,\Omega$.

Rating factors

Only grouping needs to be considered:

C_g for nine circuits $= 0.5$ (IEE Regulations Table 4C1).

Tabulated current-carrying capacity I_t

$$I_t \geq \frac{I_n}{C_g} \geq \frac{40}{0.5} \geq 80\,\text{A}$$

Cable selection

From IEE Regulations Table 4D1A, $I_t \geq 80\,\text{A} = 101\,\text{A}$ and conductor size will be 25 mm².

Voltage drop

The assumption that the whole of the design current of 32.8 A will flow in the circuit would be incorrect, as the last section will only draw:

$$\frac{32.8}{14} = 2.34 \, \text{A}$$

the section previous to that 4.68 A, the one before that 7.02 A and so on, the total volt drop being the sum of all the individual volt drops. However, this is a lengthy process and for simplicity the volt drop in this case will be based on 32.8 A over the whole length.

From IEE Regulations Table 4D1B column 3, the mV drop for a 25 mm^2 conductor is 1.8 mV.

$$\text{Volt drop} = \frac{1.8 \times 32.8 \times 45}{1000} = 2.6 \, \text{V}$$

Add this to the sub-main single-phase drop, and the total will be:

$$1.87 + 2.6 = 4.47 \, \text{V} \, (\text{acceptable})$$

Shock risk constraints

$$Z_s = Z_e + \frac{(R_1 + R_2) \times L \times 1.2}{1000}$$

In this case, Z_e will be the Z_s value for the distribution circuit.

Rearranging as before, to establish a minimum cpc size, we get:

$$R_2 = \frac{\left[(Z_s - Z_e) \times 1000 \right]}{L \times 1.2} - R_1 \left(\text{for } 25 \, \text{mm}^2 \right)$$

$$= \frac{(0.57 - 0.274) \times 1000}{45 \times 1.2} - 0.727 = 4.75 \, \text{m}\Omega$$

Therefore, the nearest value below this gives a size of 4.0 mm^2:

$$\text{Total } Z_s = 0.274 + \frac{(0.727 + 4.61)}{1000} \times 45 \times 12 = 0.56 \, \Omega$$

$$(\text{less than the maximum of } 0.57 \, \Omega)$$

Thermal constraints

$$I = \frac{U_o}{Z_s} = \frac{230}{0.56} = 410\,\text{A}$$

t from Type C CB curve for 32 A is less than 0.1 s. $k = 115$ (IEE Regulations Table 54.3)

$$s = \frac{\sqrt{410^2 \times 0.1}}{115} = 1.12\,\text{mm}^2$$

Hence our 4.0 mm² cpc is adequate. So, the calculated cable design details are as follows:

- Distribution circuit protection. 100 A BS 88 fuses, distribution circuit cable 25 mm² four-core SWA with armour as the cpc.

- Final circuit protection. 32 A Type C, BS EN 60898 MCB, final circuit cable 25 mm² singles with 4.0 mm² cpc.

Conduit and trunking sizes

Part of the design procedure is to select the correct size of conduit or trunking. The basic requirement is that the space factor is not exceeded and, in the case of conduit, that cables can be easily drawn in without damage.

For trunking, the requirement is that the space occupied by conductors should not exceed 45% of the internal trunking area. The IEE Regulations Guidance Notes/On-Site Guide give a series of tables which enable the designer to select appropriate sizes by the application of conductor/conduit/trunking terms. This is best illustrated by the following examples.

Example 1.5

What size of straight conduit 2.5 m long would be needed to accommodate ten 2.5 mm² and five 1.5 mm² stranded conductors?

Tabulated cable term for 1.5 mm² stranded = 31

Tabulated cable term for 2.5 mm² stranded = 43

$$31 \times 5 = 155$$
$$43 \times 10 = 430$$
$$\text{Total} = 585$$

The corresponding conduit term must be equal to or greater than the total cable term. Hence the nearest conduit term to 585 is 800, which gives a conduit size of 25 mm.

Example 1.6

How many $4.0 \, \text{mm}^2$ stranded conductors may be installed in a straight 3 m run of 25 mm conduit?

Tabulated conduit term for 25 mm = 800

Tabulated cable term for $4.0 \, \text{mm}^2 = 58$

$$\text{Number of cables} = \frac{800}{58} = 13.79$$

Hence thirteen $4.0 \, \text{mm}^2$ conductors may be installed.

Example 1.7

What size conduit 6 m long and incorporating two bends would be needed to house eight $6.0 \, \text{mm}^2$ conductors?

Tabulated cable term for $6.0 \, \text{mm}^2 = 58$

Overall cable term $= 58 \times 8 = 464$

Nearest conduit term above this is 600, giving 32 mm conduit.

Example 1.8

What size trunking would be needed to accommodate twenty-eight $10 \, \text{mm}^2$ conductors?

Tabulated cable term for $10 \, \text{mm}^2 = 36.3$

Overall cable term $= 36.3 \times 28 = 1016.4$

Nearest trunking term above this is 1037, giving 50 mm × 50 mm trunking.

Example 1.9

What size of trunking would be required to house the following conductors?

20, 1.5 mm² stranded

35, 2.5 mm² stranded

28, 4.0 mm² stranded

Tabulated cable term for 1.5 mm² = 8.1

Tabulated cable term for 2.5 mm² = 11.4

Tabulated cable term for 4.0 mm² = 15.2

Hence 8.1 × 20 = 162

11.4 × 35   = 399

15.2 × 28   = 425.6

Total   = 986.6

The nearest trunking term is 993, giving 100 mm × 225 mm trunking, but it is more likely that the more common 50 mm × 250 mm would be chosen (Figure 1.10).

Note

It is often desirable to make allowance for future additions to trunking systems, but care must be taken to ensure that extra circuits do not cause a change of grouping factor which could then de-rate the existing conductors below their original designed size.

Drawings

Having designed the installation it will be necessary to record the design details either in the form of a schedule for small installations or on drawings for the more complex installation. These drawings may be of the block, interconnection, layout, etc., type. The following figures indicate some typical drawings (see Figures 1.11 and 1.12).

Note the details of the design calculations shown in Figure 1.12, all of which is essential information for the testing and inspection procedure.

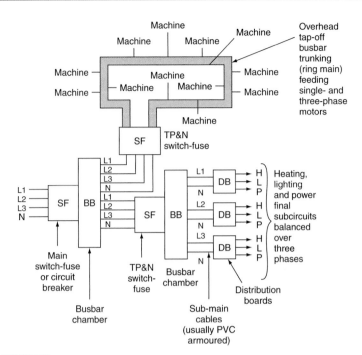

FIGURE 1.10 Layout of industrial installation.

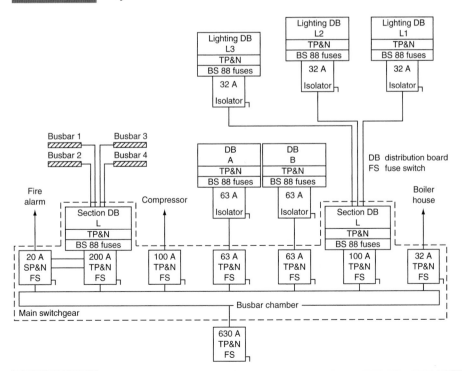

FIGURE 1.11 Distribution system, block type.

With the larger types of installation, an alphanumeric system is very useful for cross-reference between block diagrams and floor plans showing architectural symbols. Figure 1.13 shows such a system.

Distribution board 3 (DB3) under the stairs would have appeared on a diagram such as Figure 1.13, with its final circuits indicated. The floor

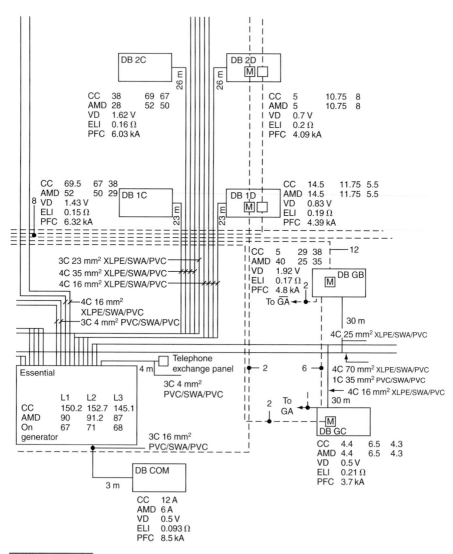

FIGURE 1.12 Distribution system, interconnection type. CC: circuit current; AMD: assumed maximum demand; VD: volt drop; ELI: earth loop impedance; PFC: prospective fault or short-circuit current.

plan shows which circuits are fed from DB3, and the number and phase colour of the protection. For example, the fluorescent lighting in the main entrance hall is fed from fuse or MCB 1 on the brown phase of DB3, and is therefore marked DB3/Br1. Similarly, the water heater circuit in the female toilets is fed from fuse or MCB 2 on the black phase, i.e. DB3/Bk2.

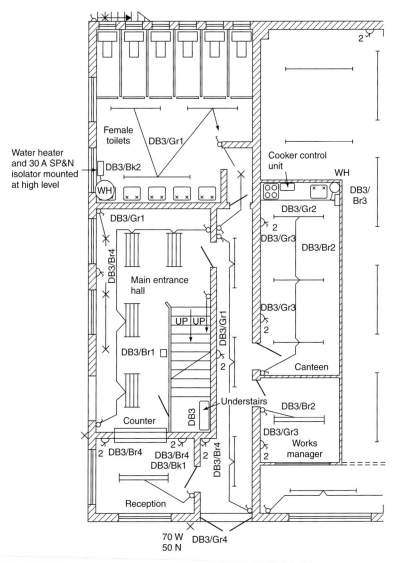

FIGURE 1.13 Example floor plan.

Figures 1.14–1.16 illustrate a simple but complete scheme for a small garage/workshop. Figure 1.14 is an isometric drawing of the garage and the installation, from which direct measurements for materials may be taken. Figure 1.15 is the associated floor plan, which cross-references with the DB schedule and interconnection details shown in Figure 1.16.

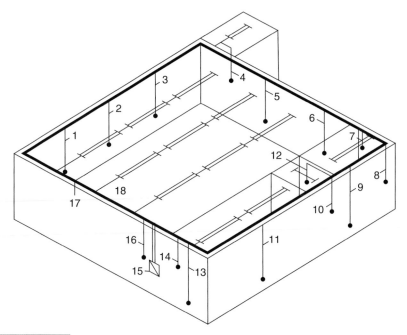

FIGURE 1.14 Isometric drawing for garage/workshop. 1 – Three-phase supply to ramp: 20 mm conduit, 2 – single-phase supply to double sockets: 20 mm conduit. Also 3, 4, 5, 6, 9, 11, 13 – single-phase supply to light switch in store: 20 mm conduit, 7 – single-phase supply to light switch in compressor: 20 mm conduit, 8 – three-phase supply to compressor: 20 mm conduit, 10 – single-phase supply to heater in WC: 20 mm conduit, 12 – single-phase supply to light switch in WC: 20 mm conduit, 14 – single-phase supply to light switch in office: 20 mm conduit, 15 – main intake position, 16 – single-phase supplies to switches for workshop lights: 20 mm conduit, 17 – 50 mm × 50 mm steel trunking, 18 – supplies to fluorescent fittings: 20 mm conduit.

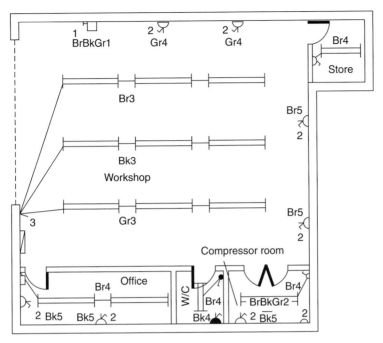

FIGURE 1.15 Floor plan for garage/workshop.

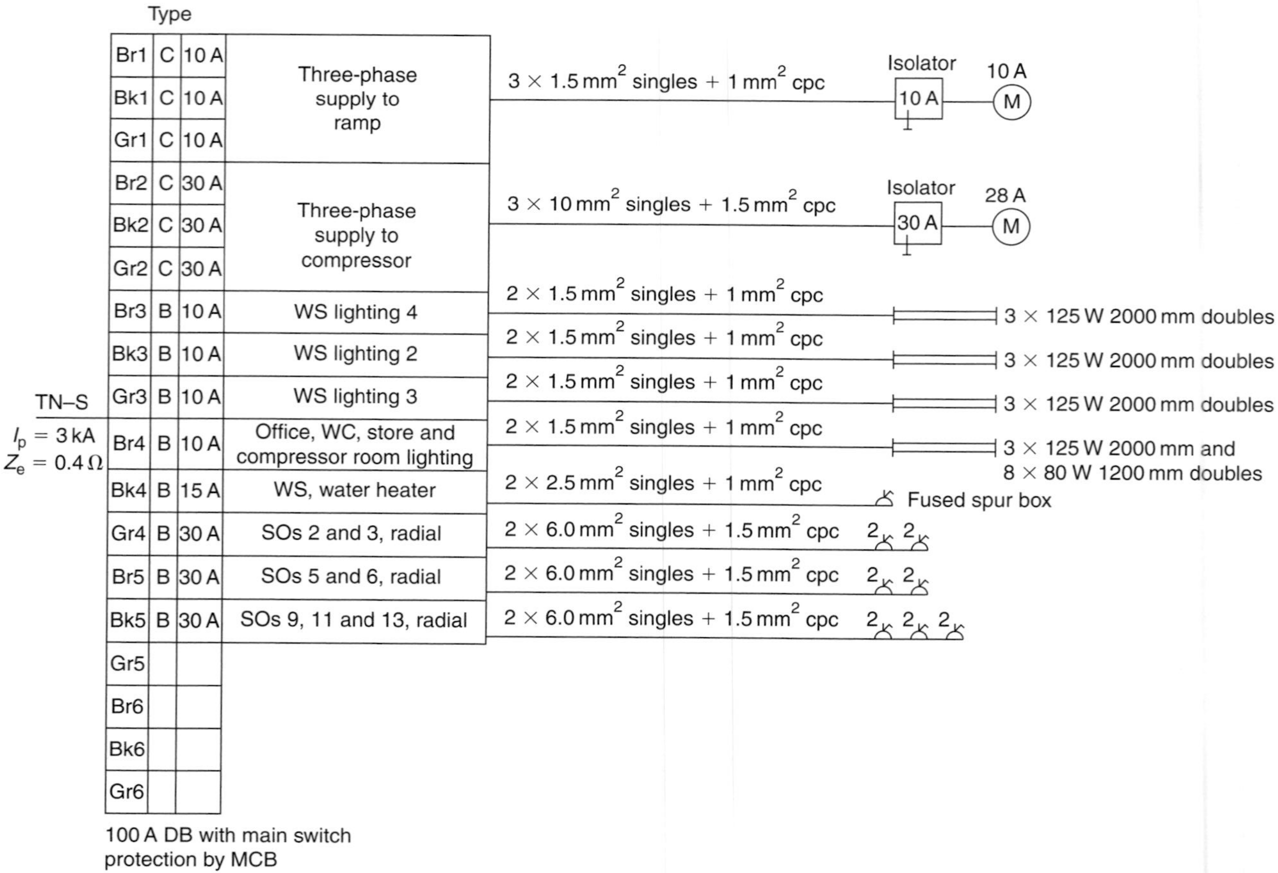

Type					
Br1	C	10 A	Three-phase supply to ramp	$3 \times 1.5\,mm^2$ singles $+ 1\,mm^2$ cpc	Isolator 10 A — 10 A — (M) 10 A
Bk1	C	10 A			
Gr1	C	10 A			
Br2	C	30 A	Three-phase supply to compressor	$3 \times 10\,mm^2$ singles $+ 1.5\,mm^2$ cpc	Isolator 28 A — 30 A — (M)
Bk2	C	30 A			
Gr2	C	30 A			
Br3	B	10 A	WS lighting 4	$2 \times 1.5\,mm^2$ singles $+ 1\,mm^2$ cpc	$3 \times 125\,W$ 2000 mm doubles
Bk3	B	10 A	WS lighting 2	$2 \times 1.5\,mm^2$ singles $+ 1\,mm^2$ cpc	$3 \times 125\,W$ 2000 mm doubles
Gr3	B	10 A	WS lighting 3	$2 \times 1.5\,mm^2$ singles $+ 1\,mm^2$ cpc	$3 \times 125\,W$ 2000 mm doubles
Br4	B	10 A	Office, WC, store and compressor room lighting	$2 \times 1.5\,mm^2$ singles $+ 1\,mm^2$ cpc	$3 \times 125\,W$ 2000 mm doubles
Bk4	B	15 A	WS, water heater	$2 \times 2.5\,mm^2$ singles $+ 1\,mm^2$ cpc	$3 \times 125\,W$ 2000 mm and $8 \times 80\,W$ 1200 mm doubles
Gr4	B	30 A	SOs 2 and 3, radial	$2 \times 6.0\,mm^2$ singles $+ 1.5\,mm^2$ cpc	Fused spur box 2 2
Br5	B	30 A	SOs 5 and 6, radial	$2 \times 6.0\,mm^2$ singles $+ 1.5\,mm^2$ cpc	2 2
Bk5	B	30 A	SOs 9, 11 and 13, radial	$2 \times 6.0\,mm^2$ singles $+ 1.5\,mm^2$ cpc	2 2 2
Gr5					
Br6					
Bk6					
Gr6					

TN–S

$I_p = 3\,kA$

$Z_e = 0.4\,\Omega$

100 A DB with main switch protection by MCB

FIGURE 1.16   Details of connection diagram for garage/workshop.

Questions

1. State the three categories of external influence.
2. Which of the main chapter headings in Part 3 of BS 7671 is relevant to starting currents?
3. State the most common method of providing Basic protection.
4. State the most common method of providing Fault protection.
5. What would be the IP code for an item of equipment that is subject to splashes and the ingress of small foreign bodies?
6. What is the impact code for an impact resistance of 5 joules?
7. What is the maximum disconnection time for a distribution circuit supplied by a TT system?
8. State the simple formula for calculation of the earth fault loop impedance of a circuit.
9. Excluding RCDs state the other method of providing 'additional protection'.
10. What type of overcurrent would flow if a motor became jammed?
11. What is pre-arcing let-through energy?
12. What is achieved in a circuit when the total let-through energy of a minor protective device is less the pre-arcing let-through energy of the major device?
13. What installation voltage condition would be likely to arise if several large motors were energized at the same time?
14. What installation voltage condition would be likely to arise if heavily inductive loads were switched off?
15. What switching device has fuses as part of the switching mechanism?
16. State the full formula for determining three-phase design current.
17. What is the term used to indicate the assumption that not all of an installation would be energized at any one time?
18. What would be the de-rating factor for a cable totally surrounded by thermal insulation for a length of 2 m?
19. State the formula used to determine whether a cpc is acceptable to withstand let-through energy.
20. What type of diagram would be used to display the simple sequential arrangement of equipment in an installation?

Answers

1. Environment, utilization and building.
2. Compatibility.
3. Insulation of live parts, barriers or enclosures.
4. Automatic disconnection of supply.
5. IP44.
6. IK08.
7. 1 s.

8. $Z_s = Z_e + (R_1 + R_2)$.

9. Supplementary equipotential bonding.

10. Overload current.

11. The electrical energy that a protective device lets through, under fault conditions, up to its cut off point.

12. Discrimination.

13. Undervoltage.

14. Overvoltage.

15. Fuse switch.

16. $I_b = \dfrac{P(\text{watts})}{\sqrt{3} \times V_L \times \text{Eff}\% \times \text{PF}}$

17. Diversity.

18. 0.5.

19. $S = \dfrac{\sqrt{I^2 t}}{k}$

20. Block diagram.

CHAPTER 2

Inspection and Testing

Important terms/topics covered in this chapter:

- Initial verification
- Periodic inspection and testing
- Inspection
- Approved test lamps and voltage indicators
- Care of instruments
- Tests and testing
- Certification

At the end of this chapter the reader should,

- know the differences between Initial verification and Periodic inspection and testing,
- be able to identify the correct certification documents to be completed,
- be aware of the importance of having accurate instrumentation,
- know the correct method of proving a circuit is dead and safe to work on,
- be able to list the relevant tests and the sequence in which they should be conducted,
- understand the theory behind, and the methods of, testing,
- understand and interpret test results.

INITIAL VERIFICATION

This is the subject of Part 6 of the IEE Regulations, and opens with the statement to the effect that it must be verified that *all* installations, before being put into service, comply with the Regulations, i.e. BS 7671. The author interprets the comment 'before being put into service by the user' as before being handed over to the user, not before the supply is connected. Clearly a supply is needed to conduct some of the tests.

IEE Wiring Regulations. DOI: 10.1016/B978-0-08-096914-5.10002-0

The opening statement also indicates that verification of compliance be carried out during the erection of the installation and after it has been completed. In any event, certain criteria must be observed:

1. The test procedure must not endanger persons, livestock or property.
2. Before any inspection and testing can even start, the person carrying out the verification must be in possession of all the relevant information and documentation. In fact the installation will fail to comply without such information (IEE Regulations 611.3 xvi). How, for example, can a verifier accept that the correct size conductors have been installed, without design details (IEE Regulations 611.3 iv)?

So, let us start, as they say, at the beginning. Armed with the results of the Assessment of General Characteristics, the designer's details, drawings, charts, etc., together with test instruments, the verification process may proceed.

INSPECTION

Usually referred to as a visual inspection, this part of the procedure is carried out against a checklist as detailed in the IEE Regulations, Section 611, Appendix 6 and in the Guidance Notes 3 for inspection and testing. Much of the initial inspection will involve access to enclosures housing live parts; hence, those parts of the installation being tested should be isolated from the supply.

Naturally, any defects found must be rectified before instrument tests are performed.

TESTING

This involves the use of test equipment and there are several important points to be made in this respect:

1. Electronic instruments must conform to BS 4743 and electrical instruments to BS 5458 and their use to HSE GS 38.

2. The Electricity at Work Regulations 1989 state an *absolute* require-ment in Regulation 4(4), that test equipment be maintained in good condition. Hence it is important to ensure that regular cali-bration is carried out.

3. Test leads used for measurement of voltages over 50 V should have shrouded/recessed ends and/or when one end is a probe, it should be fused and be insulated to within 2 mm to 4 mm of the tip and have finger guards.

4. Always use approved voltmeters, test lamps, etc.

Selection of test equipment

As has been mentioned, instruments must comply with the relevant British Standard, and provided they are purchased from established bona fide instrument manufacturers, this does not present a problem.

There is a range of instruments needed to carry out all the standard installation tests, and some manufacturers produce equipment with dual functions; indeed there are now single instruments capable of perform-ing all the fundamental tests. Table 2.1 indicates the basic tests and the instruments required.

Table 2.1 Tests and Instruments.

Test		Instrument Range	Type of Instrument
1	Continuity of ring final conductors	0.05–08 Ω	Low resistance ohmmeter
2	Continuity of protective conductors	2–0.005 Ω or less	Low resistance ohmmeter
3	Earth electrode resistance	Any value over about 3–4 Ω	Special ohmmeter
4	Insulation resistance	Infinity to less than 1 MΩ	High resistance ohmmeter
5	Polarity	None	Ohmmeter, bell, etc.
6	Earth fault loop impedance	0–2000 Ω	Special ohmmeter
7	Operation of RCD	5–500 mA	Special instrument
8	Prospective short-circuit current	2 A to 20 kA	Special instrument

APPROVED TEST LAMPS AND INDICATORS

Search your tool boxes: find, with little difficulty one would suspect, your 'neon screwdriver' or 'testascope'; locate a very deep pond; and drop it in!

Imagine actually allowing electric current at low voltage (50–1000 V AC) to pass through one's body in order to activate a test lamp! It only takes around 10–15 mA to cause severe electric shock, and 50 mA (1/20th of an ampere) to kill.

Apart from the fact that such a device will register any voltage from about 5 V upwards, the safety of the user depends entirely on the integrity of the current-limiting resistor in the unit. An electrician received a considerable shock when using such an instrument after his apprentice had dropped it in a sink of water, simply wiped it dry and replaced it in the tool box. The water had seeped into the device and shorted out the resistor. There are, however, some modern devices which are proximity devices and give some indication of the presence of voltage whatever the value. These should never be accepted as proof that a circuit is not energized!!

An approved test lamp should be of similar construction to that shown in Figure 2.1.

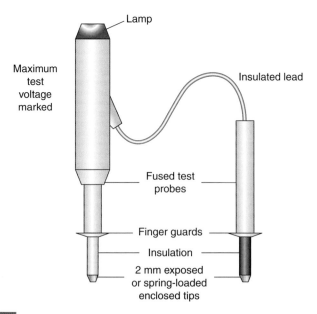

FIGURE 2.1 Approved test lamp.

The following procedure is recommended when using approved test lamps to check that live parts have been made dead:

1. Check that the test lamp is in good condition and the leads are undamaged. (This should be done regardless of the purpose of use.)
2. Establish that the lamp is sound by probing onto a known supply. This is best achieved by using a proving unit. This is simply a pocket-sized device which electronically produces 239 V DC.
3. Carry out the test to verify the circuit is dead.
4. Return to the proving unit and check the lamp again.

It has long been the practice when using a test lamp to probe between line and earth for an indication of a live supply on the line terminal. However, this can now present a problem where RCDs exist in the circuit, as of course the test is applying a deliberate line-to-earth fault.

Some test lamps have LED indicators, and the internal circuitry of such test lamps limits the current to earth to a level below that at which the RCD will operate. The same limiting effect applies to multi-meters. However, it is always best to check that the testing device will have no effect on RCDs.

CALIBRATION, ZEROING/NULLING AND CARE OF INSTRUMENTS

Precise calibration of instruments is usually well outside the province of the electrician, and would normally be carried out by the manufacturer or local service representative. A check, however, can be made by the user to determine whether calibration is necessary by comparing readings with an instrument known to be accurate, or by measurement of known values of voltage, resistance, etc. on one of the many 'check boxes' currently marketed. However, as we have already seen, regular calibration is a legal requirement.

It may be the case that readings are incorrect simply because the instrument is not zeroed/nulled before use, or because the internal battery needs replacing. Most modern instruments have battery condition indication, and of course this should never be ignored.

Always adjust any selection switches to the off position after testing. Too many instrument fuses are blown when, for example, a multi-meter is inadvertently left on the ohms range and then used to check for mains voltage.

The following set procedure may seem rather basic but should ensure trouble-free testing:

1. Check test leads for obvious defects.
2. Zero/null the instrument.
3. Select the correct range for the values anticipated. If in doubt, choose the highest range and gradually drop down.
4. Make a record of the test results.
5. When an unexpected reading occurs, make a quick check on the test leads just to ensure that they are not open circuited.
6. Return switches/selectors to the off position.
7. Replace instrument and leads in carrying case.

THE TESTS

The IEE Regulations indicate a preferred sequence of tests and state that if, due to a defect, compliance cannot be achieved, the defect should be rectified and the test sequence started from the beginning. The tests for 'Site applied insulation', 'Protection by separation', and 'Insulation of non-conducting floors and walls' all require specialist high voltage equipment and in consequence will not be discussed here. The sequence of tests for an initial inspection and test is as follows:

1. Continuity of protective conductors.
2. Continuity of ring final circuit conductors.
3. Insulation resistance.
4. Protection by barriers or enclosures.
5. Polarity.
6. Earth electrode resistance.
7. Earth fault loop impedance.
8. Additional protection (RCDs).
9. Prospective fault current (PFC) between live conductors and to earth.

10. Phase sequence.
11. Functional testing.
12. Voltage drop.

One other test not included in Part 7 of the IEE Regulations but which nevertheless has to be carried out is external earth fault loop impedance (Z_e).

CONTINUITY OF PROTECTIVE CONDUCTORS

These include the cpcs of radial circuits, main and supplementary Protective bonding conductors. Two methods are available: either can be used for cpcs, but bonding can only be tested by the second.

Method 1

At the distribution board, join together the line conductor and its associated cpc. Using a low resistance ohmmeter, test between line and cpc at all the outlets in the circuit. The reading at the farthest point will be $(R_1 + R_2)$ for that circuit. Record this value; after correction for temperature it may be compared with the designer's value (more about this later).

Method 2

Connect one test instrument lead to the main earthing terminal, and a long test lead to the earth connection at all the outlets in the circuit. Record the value after deducting the lead resistance. An idea of the length of conductor is valuable, as the resistance can be calculated and compared with the test reading. Table 2.2 gives resistance values already calculated for a range of lengths and sizes.

It should be noted that these tests are applicable only to 'all insulated' systems, as installations using metallic conduit and trunking, MICC and SWA cables will produce spurious values due to the probable parallel paths in existence. This is an example of where testing needs to be carried out during the erection process and before final connections and bonding are in place.

Table 2.2   Resistance (Ω) of Copper Conductors at 20 °C

CSA (mm²)	Length (m)									
	5	10	15	20	25	30	35	40	45	50
1.0	0.09	0.18	0.27	0.36	0.45	0.54	0.63	0.72	0.82	0.90
1.5	0.06	0.12	0.18	0.24	0.30	0.36	0.42	0.48	0.55	0.61
2.5	0.04	0.07	0.11	0.15	0.19	0.22	0.26	0.30	0.33	0.37
4.0	0.023	0.05	0.07	0.09	0.12	0.14	0.16	0.18	0.21	0.23
6.0	0.02	0.03	0.05	0.06	0.08	0.09	0.11	0.13	0.14	0.16
10.0	0.01	0.02	0.03	0.04	0.05	0.06	0.07	0.08	0.09	0.10
16.0	0.006	0.01	0.02	0.023	0.03	0.034	0.04	0.05	0.051	0.06
25.0	0.004	0.007	0.01	0.015	0.02	0.022	0.026	0.03	0.033	0.04
35.0	0.003	0.005	0.008	0.01	0.013	0.016	0.019	0.02	0.024	0.03

If conduit, trunking or SWA is used as the cpc, then the verifier has the option of first inspecting the cpc along its length for soundness then conducting the long-lead resistance test.

CONTINUITY OF RING FINAL CIRCUIT CONDUCTORS

The requirement of this test is that each conductor of the ring is continuous. It is, however, not sufficient to simply connect an ohmmeter, a bell, etc., to the ends of each conductor and obtain a reading or a sound.

So what is wrong with this procedure? A problem arises if an interconnection exists between sockets on the ring, and there is a break in the ring beyond that interconnection. From Figure 2.2 it will be seen that a simple resistance or bell test will indicate continuity via the interconnection. However, owing to the break, sockets 4–11 are supplied by the spur from socket 12, not a healthy situation. So how can one test to identify interconnections?

There are three methods of conducting such a test. Two are based on the principle that resistance changes with a change in length or CSA; the other, predominantly used, relies on the fact that the resistance measured across any diameter of a circular loop of conductor is the same. Let us now consider the first two.

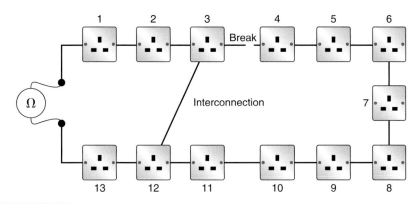

FIGURE 2.2 Ring circuit with interconnection.

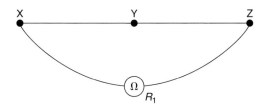

FIGURE 2.3 End to end conductor resistance.

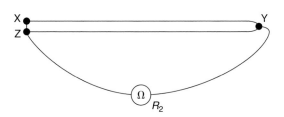

FIGURE 2.4 Doubled over end to end conductor resistance.

Method 1

If we were to take a length of conductor XYZ and measure the resistance between its ends (Figure 2.3), then double it over at Y, join X and Z, and measure the resistance between XZ and Y (Figure 2.4), we would find that the value was approximately a quarter of the original. This is because the length of the conductor is halved and hence so is the resistance, and the CSA is doubled and so the resistance is halved again.

In order to apply this principle to a ring final circuit, it is necessary to know the position of the socket nearest the mid-point of the ring.

The test procedure is then as follows for each of the conductors of the ring:

1. Measure the resistance of the ring conductor under test between its ends before completing the ring in the fuse board. Record this value, say R_1.
2. Complete the ring.
3. Using long test leads, measure between the completed ends and the corresponding terminal at the socket nearest the mid-point of the ring. Record this value, say R_2. (The completed ends correspond to point XZ in Figure 2.4, and the mid-point to Y.)
4. Measure the resistance of the test leads, say R_3, and subtract this value from R_2, i.e. $R_2 - R_3 = R_4$ say.
5. A comparison between R_1 and R_4 should reveal, if the ring is healthy, that R_4 is approximately a quarter of R_1.

Method 2

The second method tests two ring circuit conductors at once, and is based on the following.

Take two conductors XYZ and ABC and measure their resistances (Figure 2.5). Then double them both over, join the ends XZ and AC and the mid-points YB, and measure the resistance between XZ and AC (Figure 2.6). This value should be a quarter of that for XYZ plus a quarter of that for ABC.

If both conductors are of the same length and CSA, the resultant value would be half that for either of the original resistances.

Applied to a ring final circuit, the test procedure is as follows:

1. Measure the resistance of both line and neutral conductors before completion of the ring. They should both be the same value, say R_1.
2. Complete the ring for both conductors, and bridge together line and neutral at the mid-point socket (this corresponds to point YB in Figure 2.6). Now measure between the completed line and neutral ends in the fuse board (points XZ and AC in Figure 2.6). Record this value, say R_2.

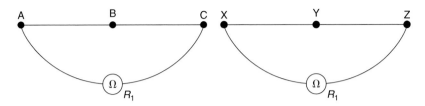

FIGURE 2.5   End to end conductor resistance.

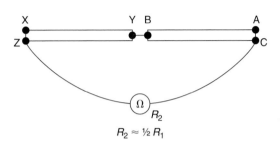

FIGURE 2.6   Doubled over conductors in parallel.

3. R_2 should be, for a healthy ring, approximately half of R_1 for either line or neutral conductor. When testing the continuity of a cpc which is a different size from either line or neutral, the resulting value R_2 should be a quarter of R_1 for line or neutral plus a quarter of R_1 for the cpc.

Method 3 (generally used)

The third method is based on the measurement of resistance at any point across the diameter of a circular loop of conductor (Figure 2.7).

As long as the measurement is made across the diameter of the ring, all values will be the same. The loop of conductor is formed by crossing over and joining the ends of the ring circuit conductors at the fuse board. The test is conducted as follows:

1. Identify both 'legs' of the ring.
2. Join one line and one neutral conductor of opposite legs of the ring.
3. Obtain a resistance reading between the other line and neutral (Figure 2.8). (A record of this value is important.)
4. Join these last two conductors (Figure 2.9).

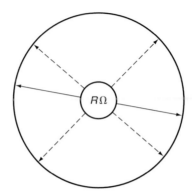

FIGURE 2.7 Resistance across diameter of circle of conductor.

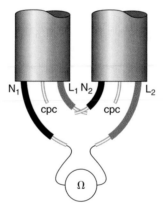

FIGURE 2.8 End to end double loop resistance.

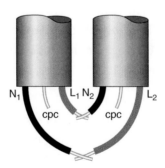

FIGURE 2.9 Both ends cross connected.

5. Measure the resistance value between L and N at each socket on the ring. All values should be the same, approximately a quarter of the reading in (3) above.

The test is now repeated but the neutral conductors are replaced by the cpcs. If the cable is twin with cpc, the cpc size will be smaller than the line conductor, and although the readings at each socket will be substantially the same, there will be a slight increase in values towards the centre of the ring, decreasing back towards the start. The highest reading represents $R_1 + R_2$ for the ring.

The basic principle of this method is that the resistance measured between any two points, equidistant around a closed loop of conductor, will be the same.

Such a loop is formed by the line and neutral conductors of a ring final circuit (Figure 2.10).

Let the resistance of conductors be as shown.

R measured between L and N on socket A will be:

$$\frac{0.2+0.5+0.2+0.3+0.4+0.1+0.3}{2} = \frac{2}{2} = 1\Omega$$

R measured between L and N at socket B will be:

$$\frac{0.3+0.2+0.5+0.2+0.3+0.4+0.1}{2} = \frac{2}{2} = 1\Omega$$

Hence all sockets on the ring will give a reading of 1Ω between L and N.

If there were a break in the ring in, say, the neutral conductor, all measurements would have been 2, incorrectly indicating to the tester that the ring was continuous. Hence the relevance of step 3 in the test procedure, which at least indicates that there is a continuous L–N loop, even if an interconnection exists. Figure 2.11 shows a healthy ring with interconnection.

Here is an example that shows the slight difference between measurements on the line/cpc test. Consider a 30 m ring final circuit wired in 2.5 mm^2 with a 1.5 mm^2 cpc. Figure 2.12 illustrates this arrangement when cross-connected for test purposes.

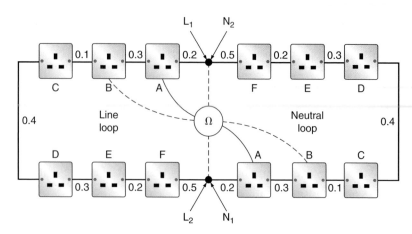

FIGURE 2.10 Equidistant loop measurement.

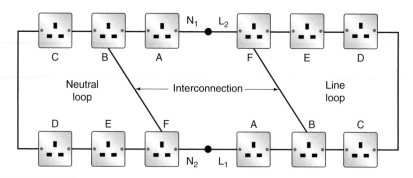

FIGURE 2.11 Healthy ring circuit with an interconnection.

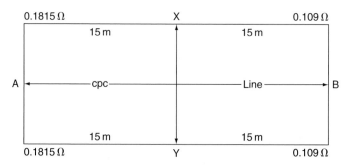

FIGURE 2.12 Ring with 2.5mm² line and 1.5mm² cpc.

From the resistance tables, $1.5\,mm^2$ conductor is seen to have a resistance of $12.1\,m\Omega/m$, and $2.5\,mm^2$, $7.27\,m\Omega/m$. This gives the resistance from X to A as $15 \times 12.1/1000 = 0.1815\,\Omega$ and from X to B as $15 \times 7.27/1000 = 0.109$. The same values apply from Y to A and Y to B.

So measuring across X and Y we have $2 \times 0.1815 = 0.363$, in parallel with $2 \times 0.109 = [(0.363 \times 0.218)/(0.363 + 0.218)]\,\Omega$ (product over sum) $= 0.137\,\Omega$.

Measuring across A and B (the mid-point) gives $0.1815 + 0.109 = 0.29\,\Omega$, in parallel with the same value, i.e. $0.29\,\Omega$, which gives $0.29/2 = 0.145\,\Omega$.

While there is a difference of $0.008\,\Omega$ the amount is too small to suggest any faults on the ring.

Note: If the line–neutral and line–cpc tests prove satisfactory this is also an indication that the polarity at each socket outlet is correct.

Protection by barriers or enclosures

If an enclosure/barrier is used to house or obscure live parts, and is not a factory-built assembly, it must be ascertained whether or not it complies with the requirements of the IP codes IP2X or IPXXB, or IP4X or IPXXD. For IP2X or IPXXB, the test is made using the British Standard Finger, which is connected in series with a lamp and a supply of not less than 40 V and not more than 50 V. The test finger is pushed into or behind the enclosure/barrier and the lamp should not light (Figure 2.13).

The test for IP4X or IPXXD is made with a 1.0 mm diameter wire with its end cut at right angles to its length. The wire should not enter the enclosure to conform to IP4X. It may enter for 100 mm without touching live parts to conform to IPXXD.

INSULATION RESISTANCE

An insulation resistance tester, which is a high resistance ohmmeter, is used for this test. The test voltages and minimum $M\Omega$ values are shown in Table 2.3.

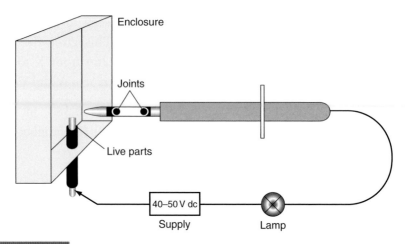

Enclosure

Joints

Live parts

40–50 V dc

Supply

Lamp

FIGURE 2.13 Penetration test.

Table 2.3 Test Voltage and Minimum MΩ Values for Insulation Resistance Testing

ELV Circuits (SELV and PELV)		LV Circuits up to 500 V		LV Circuits Above 500 V	
V	M	V	M	V	M
250	0.5	500	1	1000	1

Clearly with voltages of these levels, there are certain precautions to be taken prior to the test being carried out. Persons should be warned, and sensitive electronic equipment disconnected or unplugged. A common example of this is the dimmer switch. Also, as many accessories have indicator lamps, and items of equipment such as fluorescent fittings have capacitors fitted, these should be disconnected as they will give rise to false readings.

The test procedure is as follows:

Poles to earth (Figure 2.14)
1. Isolate supply.
2. Ensure that all protective devices are in place and all switches are closed.
3. Link all poles of the supply together (where appropriate).
4. Test between the linked poles and earth.

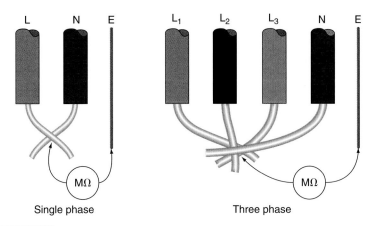

FIGURE 2.14   Test between live conductors and earth.

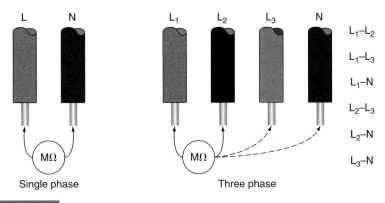

FIGURE 2.15   Test between live conductors.

Between poles (Figure 2.15)

1. As previous test.
2. As previous test.
3. Remove all lamps, equipment, etc.
4. Test between poles.

Test results on disconnected equipment should conform to the relevant British Standard for that equipment. In the absence of a British Standard, the minimum value is $0.5\,M\Omega$.

For small installations, the tests are performed on the whole system, whereas for larger complex types, the installation may be sub-divided into sections and tests performed on each section. The reason for this is that as conductor insulation and the circuits they supply are all in parallel, a test on the whole of a large installation would produce pessimistically low readings even though no faults exist.

Although for standard 400 V/230 V installations the minimum value of insulation resistance is 1 MΩ, a reading of less than 2 MΩ should give rise to some concern. Circuits should be tested individually to locate the source/s of such a low reading.

POLARITY

It is required that all fuses and single-pole devices such as single-pole circuit breakers and switches are connected in the line conductor only. It is further required that the centre contact of Edison screw lampholders be connected to the line conductor (BS EN 60298 E14 and E27 ES types are exempt as the screwed part is insulated) and that socket outlets and similar accessories are correctly connected.

RING FINAL CIRCUITS

If method 3 for testing ring circuit conductor continuity was performed, then any cross-polarity would have shown itself and been rectified. Hence no further test is necessary. However, if method 1 or 2 were used, and the mid-point socket was correct, reversals elsewhere in the ring would not be detected and therefore two tests are needed:

1. Link completed line and cpc loops together at the fuse board and test between L and E at each socket. A no reading result will indicate a reversed polarity (Figure 2.16).
2. Repeat as in 1, but with L and N linked.

RADIAL CIRCUITS

For radial circuits, the test method 1 for continuity of protective conductors will have already proved correct polarity. It just remains to check

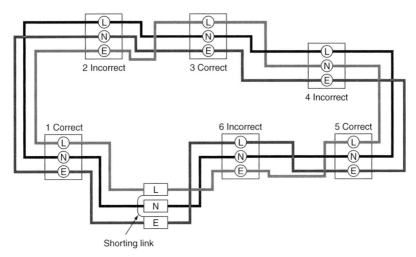

FIGURE 2.16 Polarity test if done separately to method 3 ring test.

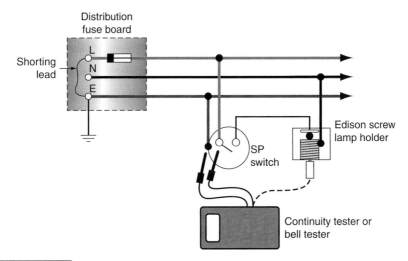

FIGURE 2.17 Polarity on ES lampholder.

the integrity of the neutral conductor for socket outlet circuits, and that switch wires and neutrals are not mixed at lighting points. This is done by linking L and N at the fuse board and testing between L and N at each outlet and between N and switch wire at each lighting point.

Also for lighting circuits, to test for switches in line conductors, etc., link L and E at the fuse board and test as shown in Figure 2.17.

EARTH ELECTRODE RESISTANCE

If we were to place an electrode in the earth and measure the resistance between the electrode and points at increasingly larger distances from it, we would notice that the resistance increased with distance until a point was reached (usually around 2.5 m) beyond which no increase in resistance was noticed (Figure 2.18).

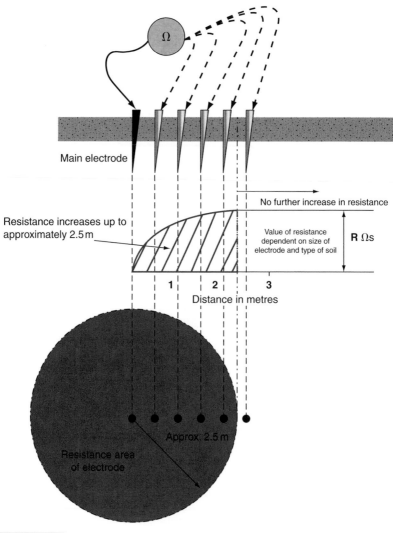

FIGURE 2.18 Electrode resistance areas.

It is a requirement of the Regulations that for a TT system, exposed conductive parts must be connected via protective conductors to an earth electrode, and that the protection is by either an RCD or an overcurrent device, the RCD being preferred. Conditional on this is the requirement that the product of the sum of the resistances of the earth electrode and protective conductors, and the operating current of the protective device, shall not exceed 50 V, i.e. $R_a \times I_a \leq 50$ V.

(R_a is the sum of the resistances of the earth electrode and the protective conductors connecting it to the exposed conductive part.)

Clearly then, there is a need to measure the resistance of the earth electrode. This may be done in either of two ways.

Method 1

Based on the principle of the potential divider (Figure 2.19, page 76), an earth resistance tester is used together with test and auxiliary electrodes spaced as shown in Figure 2.20. This spacing ensures that resistance areas do not overlap.

The method of test is as follows:

1. Place the current electrode (C2) away from the electrode under test, approximately 10 times its length, i.e. 30 m for a 3 m rod.
2. Place the potential electrode mid-way.
3. Connect test instrument as shown.

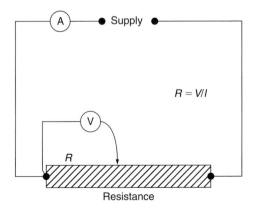

FIGURE 2.19 Potential divider.

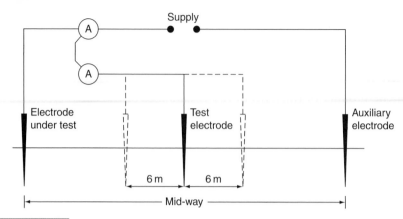

FIGURE 2.20 Positions of test electrodes.

4. Record resistance value.
5. Move the potential electrode approximately 6 m either side of the mid-position, and record these two readings.
6. Take an average of these three readings (this is the earth electrode resistance).

Three readings obtained from an earth electrode resistance test were 181 Ω, 185 Ω and 179 Ω. What is the value of the electrode resistance?

$$\text{Average value} = \frac{181 + 185 + 179}{3} = 181.67\,\Omega$$

For TT systems the result of this test will indicate compliance if the product of the electrode resistance and the operating current of the overcurrent device does not exceed 50 V.

Method 2

On TT systems protected by an RCD, a loop impedance tester is used and effectively measures Z_e, which is taken as the earth electrode resistance.

EXTERNAL LOOP IMPEDANCE Z_e

This is carried out by connecting an earth fault loop impedance tester between the supply line conductor and the earthing conductor at the intake position with the earthing conductors disconnected. This ensures that parallel resistance paths will not affect the reading. Wherever possible the installation should be isolated from the supply during the test. If this is not possible then all circuits should be isolated. When the test is completed reconnect the earthing conductor.

EARTH FAULT LOOP IMPEDANCE Z_s

This has to be measured in order to ensure that protective devices will operate in the specified time under fault conditions. As the value of $(R_1 + R_2)\Omega$ for a particular circuit will have already been established, Z_s may be found by simply adding the $R_1 + R_2$ value to Z_e. Alternatively, it may be measured directly at the extremity of a particular circuit. Whichever method is used, the value obtained will need to be corrected to compensate for ambient and conductor operating temperatures before a comparison is made with the tabulated values of Z_s in the Regulations.

Note

All main protective and supplementary bonding must be in place during this test.

ADDITIONAL PROTECTION

Residual current devices

Only the basic type of RCD will be considered here. Clearly, such devices must operate to their specification; an RCD tester will establish this. As with loop impedance testing, care must be taken when conducting this test as an intentional earth fault is created in the installation. In consequence, a loop impedance test must be conducted first to confirm that an earth path exists or the RCD test could prove dangerous.

It is important to know why an RCD has been installed as this has direct effect on the tests performed. The tests are as follows:

1. With the tester set to the RCD rating, half the rated current is passed through the device. It should not trip.
2. With full rated current passed through the device, it should trip within 200 ms (300 ms for RCBOs).

3. For RCDs having a residual current rating of 30 mA or less, a test current of $5 \times I_{\Delta n}$ should be applied and the device should operate in 40 ms or less.

4. All RCDs have a test button which should be operated to ensure the integrity of the tripping mechanism. It does not check any part of the earthing arrangements or the device's sensitivity. As part of the visual inspection, it should be verified that a notice, indicating that the device should be tested via the test button quarterly, is on or adjacent to the RCD.

There seems to be a popular misconception regarding the ratings and uses of RCDs in that they are the panacea for all electrical ills and the only useful rating is 30 mA!

Firstly, RCDs are not fail-safe devices; they are electromechanical in operation and can malfunction. Secondly, general purpose RCDs are manufactured in ratings from 5 mA to 1000 mA (30 mA, 100 mA, 300 mA and 500 mA being the most popular) and have many uses. The following list indicates residual current ratings and uses as mentioned in BS 7671.

Requirements for RCD protection

30 mA

- All socket outlets rated at not more than 20 A and for un-supervised general use.
- Mobile equipment rated at not more than 32 A for use outdoors.
- All circuits in a bath/shower room.
- Preferred for all circuits in a TT system.
- All cables installed less than 50 mm from the surface of a wall or partition (in the safe zones) if the installation is un-supervised, and also at any depth if the construction of the wall or partition includes metallic parts.
- In zones 0, 1 and 2 of swimming pool locations.
- All circuits in a location containing saunas, etc.
- Socket outlet final circuits not exceeding 32 A in agricultural locations.
- Circuits supplying Class II equipment in restrictive conductive locations.

- Each socket outlet in caravan parks and marinas and final circuit for houseboats.
- All socket outlet circuits rated not more than 32 A for show stands, etc.
- All socket outlet circuits rated not more than 32 A for construction sites (where reduced low voltage, etc. is not used).
- All socket outlets supplying equipment outside mobile or transportable units.
- All circuits in caravans.
- All circuits in circuses, etc.
- A circuit supplying Class II heating equipment for floor and ceiling heating systems.

500 mA

- Any circuit supplying one or more socket outlets of rating exceeding 32 A, on a construction site.

300 mA

- At the origin of a temporary supply to circuses, etc.
- Where there is a risk of fire due to storage of combustible materials.
- All circuits (except socket outlets) in agricultural locations.

100 mA

- Socket outlets of rating exceeding 32 A in agricultural locations.

Where loop impedance values cannot be met, RCDs of an appropriate rating can be installed. Their rating can be determined from

$$I_{\Delta n} = 50 / Z_s$$

where $I_{\Delta n}$ is the rated operating current of the device, 50 is the touch voltage and Z_s is the measured loop impedance.

Prospective fault current

A PFC tester, usually incorporated with a loop impedance tester, is used for this. When testing at the intake position, probes and/or clips will be needed and hence great care needs to be taken when connecting to live terminals, etc. Measurements are taken between L and N, and L and E and the highest value recorded. For three phase supplies, the line to line PFC is determined from multiplying the L to N reading by $\sqrt{3}$ (1.732) or more simply by 2.

Phase sequence

For multi-phase circuits, e.g. supplies to three-phase motors, etc, it is important to check that the phase sequence is correct to ensure correct direction of rotation. A phase sequence instrument is used which is basically a small three-phase motor.

Functional testing

Tests on assemblies

These are carried out on a switchgear, interlock, control gear, etc., to ensure that they are mounted and installed according to the Requirements of the 17th Edition.

Voltage drop

There may be occasions when verification of voltage drop is required. This would be achieved by calculation or by reference to charts or tables.

Periodic inspection and testing

After an installation has had an initial verification and been put into service, there is a requirement for regular periodic verification to take place. In some cases where, for example, a Local Authority is involved, the interval between tests is mandatory. In other cases the interval is only a recommendation. For example, the recommended time between tests on domestic installations is 10 years, whereas places of public entertainment have a mandatory interval of one year.

Clearly, periodic tests may prove difficult, as premises are usually occupied and in full service, and hence careful planning and consultation are needed in order to minimize any disruption. A thorough visual inspection should be undertaken first, as this will indicate to the experienced inspector the depth to which he or she needs to go with the instrument tests, and an even more rigorous investigation may be required if drawings/design data are not available.

The visual inspection will need to take into account such items as safety, wear and tear, corrosion, signs of overloading, mechanical damage, etc. In many instances, a sample of items inspected may be taken, for example a minimum of 10% of switching devices may be taken. If, however, the sample indicates considerable deterioration then all items must be inspected.

The test sequence where relevant and where possible should be the same as that for an initial verification. This is not essential: As with visual inspection, sample tests may be made, usually 10%, with the proviso that this is increased in the event of faults being found.

In the light of previous comments regarding sampling, it is clear that periodic verification is subjective, varying from installation to installation. It is also more dangerous and difficult and hence requires the inspector to have considerable experience. Accurate and coherent records must be made and given to the person/s ordering the work. Such records/reports must indicate any departures from or non-compliances with the Regulations, any restrictions in the testing procedure, any dangerous situations, etc.; if the installation was erected according to an earlier edition of the Regulations, it should be tested as far as possible to the requirements of the 17th Edition, and a note made to this effect on the test report.

It should be noted that if an installation is effectively supervised in normal use, then Periodic Inspection and Testing can be replaced by regular maintenance by skilled persons. This would only apply to, say, factory installations where there are permanent maintenance staff.

Certification

For new installations, alterations and additions, an Electrical Installation Certificate (EIC) is issued.

For changes that do not include the provision of a new circuit a Minor Electrical Installation Works Certificate (MEIWC) is issued. For existing installations a Condition report (CR) is issued.

Schedules of inspections and schedules of test results **must** accompany both EIC's and CR's. Without these schedules, the Certificates are **not** valid.

Questions

1. What type of Inspection and Testing should be carried out for the installation of a new circuit in an existing installation?
2. What documentation should be completed after an Inspection and Test on an installation requested by an insurance company?
3. What statutory document suggests that test instruments should be regularly calibrated?
4. State the three stages, using an approved voltage indicator, required to prove that a circuit is dead and safe to work on.
5. List, in sequence, the 'dead' tests to be carried out on a newly installed shower circuit.
6. What would be the approximate resistance of a 15 m length of 10 mm^2 copper conductor?
7. What other test is automatically conducted when carrying out a ring circuit continuity test?
8. Below what value of insulation resistance is further investigation considered necessary?
9. Why are BS EN 60298 E14 and E27 Edison screw lampholders exempt from polarity tests?
10. What is the value of earth electrode resistance if the three test values are 172 Ω, 174 Ω and 170 Ω?
11. Why should a test for external loop impedance Z_e require disconnection of the earthing conductor?
12. Why should an earth fault loop impedance test be conducted before an RCD test?
13. What would be the approximate value of the line to line PFC in three phase installation if the line to neutral value was found to be 3 kA?
14. What would be the consequence of incorrect phase sequence on a three phase motor?
15. How often is it recommended that an RCD is tested via its integral test button?

Answers

1. Initial verification.
2. Condition report, and schedules of inspections and test results.
3. EAWR 1989.
4. Prove the instrument, test that the circuit is dead, re-prove the instrument.
5. Continuity of protective conductors, insulation resistance, polarity.
6. $0.03\,\Omega$.
7. Polarity.
8. $2\,M\Omega$.
9. The screw threads are made of an insulation material.
10. $172\,\Omega$.
11. To avoid measuring parallel paths.
12. To ensure that a good earth return path is present. Without one the RCD test could present a shock risk as the test puts an earth fault on the installation and the RCD may not trip.
13. $6\,kA$.
14. A reversal of rotation.
15. Quarterly.

CHAPTER 3

Special Locations IEE Regulations Part 7

Important terms/topics covered in this chapter:

- Zonal requirements for bathrooms, etc., swimming pools, etc., and saunas
- External influences
- IP and IK codes
- Wiring systems

At the end of this chapter the reader should,

- know why particular locations are termed 'Special',
- have a broad understanding of the important points in each location,
- know the relevant codes for protection against water, foreign bodies and impact.

INTRODUCTION

The bulk of BS 7671 relates to typical, single- and three-phase, installations. There are, however, some special installations or locations that have particular requirements. Such locations may present the user/occupant with an increased risk of death or injuries from electric shock.

BS 7671 categorizes these special locations in Part 7 and they comprise the following:

Section 701	Bathrooms, shower rooms, etc.
Section 702	Swimming pools and other basins
Section 703	Rooms containing sauna heaters
Section 704	Construction and demolition sites
Section 705	Agricultural and horticultural premises
Section 706	Conducting locations with restrictive movement
Section 708	Caravan/camping parks

IEE Wiring Regulations. DOI: 10.1016/B978-0-08-096914-5.10003-2

Section 709	Marinas and similar locations
Section 710	Medical locations
Section 711	Exhibitions, shows and stands
Section 712	Solar photovoltaic (PV) power supply systems
Section 717	Mobile or transportable units
Section 721	Caravans and motor caravans
Section 729	Operating and maintenance gangways
Section 740	Amusement devices, fairgrounds, circuses, etc.
Section 753	Floor and ceiling heating systems

Let us now briefly investigate the main requirements for each of these special locations.

BS 7671 SECTION 701: BATHROOMS, ETC.

This section deals with rooms containing baths, shower basins or areas where showers exist but with tiled floors, for example leisure/recreational centres, sports complexes, etc.

Each of these locations is divided into zones 0, 1 and 2, which give an indication of their extent and the equipment/wiring, etc. that can be installed in order to reduce the risk of electric shock.

So! Out with the tape measure, only to find that in a one-bedroom flat, there may be no zone 2. How can you conform to BS 7671?

The stark answer (mine) is that you may not be able to conform exactly. You do the very best you can in each particular circumstance to ensure safety. Let us not forget that the requirements of BS 7671 are based on reasonableness.

Zone 0

This is the interior of the bath tub or shower basin or, in the case of a shower area without a tray, it is the space having a depth of 100 mm above the floor out to a radius of 600 mm from a fixed water outlet or 1200 mm radius for an extended zone 1 (Figure 3.1).

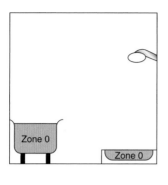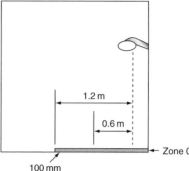

FIGURE 3.1 Zone 0.

Note

- Only SELV (12 V) or ripple-free DC may be used as a measure against electric shock, the safety source being outside zones 0, 1 and 2.
- Other than current using equipment specifically designed for use in this zone, **no** switchgear or accessories are permitted.
- Equipment designed for use in this zone must be to at least IPX7.
- Only wiring associated with equipment in this zone may be installed.

Zone 1

This extends above zone 0 around the perimeter of the bath or shower basin to 2.25 m above the floor level, and includes any space below the bath or basin that is accessible without the use of a key or tool. For showers without basins, zone 1 extends out to a radius of 600 mm from a fixed water outlet or 1200 mm radius for an extended zone 1. (Figures 3.2 and 3.3).

Note

- Other than switches and controls of equipment specifically designed for use in this zone, and cord operated switches, only SELV switches are permitted.
- Provided they are suitable, fixed items of current using equipment such as:
  Showers
  Shower pumps
  Towel rails
  Luminaires, etc.
- Equipment designed for use in this zone must be to at least IPX4, or IPX5, where water jets are likely to be used for cleaning purposes.

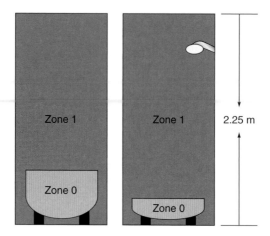

FIGURE 3.2   Zones 0 and 1.

Zone 2

This extends 600 mm beyond zone 1 and to a height of 2.25 m above floor level (Figure 3.3).

> ### Note
>
> - Other than switches and controls of equipment specifically designed for use in this zone, and cord operated switches, only SELV switches are permitted.
> - Equipment designed for use in this zone must be to at least IPX4, or IPX5 where water jets are likely to be used for cleaning purposes.
> - For showers without basins there is no zone 2, just an extended zone 1.
> - Socket outlets other than SELV may not be installed within 3 m of the boundary of zone 1.

Additional protection

All low voltage circuits of these locations must be protected by a 30 mA RCD, or less, that operates within 40 ms at $5 \times I_{\Delta n}$.

Supplementary equipotential bonding

Supplementary bonding may be established connecting together the cpcs, exposed and extraneous conductive parts within the location.

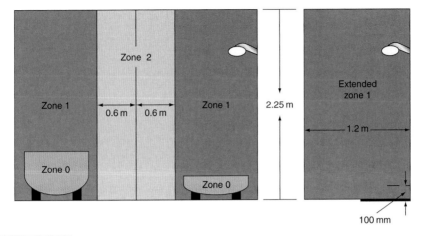

FIGURE 3.3   Zones 0, 1 and 2.

Such extraneous conductive parts will include:

- metallic gas, water, waste and central heating pipes
- metallic structural parts that are accessible to touch
- metal baths and shower basins.

This bonding may be carried out inside or outside the location, preferably close to the entry of the extraneous conductive parts to the location.

However, this bonding may be omitted if the premises has a protective earthing and automatic disconnection system in place; all extraneous conductive parts of the locations are connected to the protective bonding and all circuits are residual current device (RCD) protected (which they have to be anyway!!).

Electric floor units may be installed below any zone provided that they are covered with an earthed metal grid or metallic sheath and connected to the protective conductor of the supply circuit.

BS 7671 SECTION 702: SWIMMING POOLS

In a similar fashion to bathrooms and shower rooms, etc., swimming pool locations are also divided into zones 0, 1 and 2:

Zone 0 is in the pool/basin or fountain.

Zone 1 extends 2.0 m horizontally from the rim of zone 0 and 2.5 m vertically above it regardless of the pool being above or below ground level. If there are diving boards, shutes or viewing galleries, etc. the height extends to a point 2.5 m from their top surface and 1.5 m horizontally either side of such shutes, etc.

Zone 2 extends a further 1.5 m horizontally from the edge of zone 1 and 2.5 m above ground level.

Now, what can we install in these zones?

Zones 0 and 1

Protection against shock

Only SELV to be used.

Wiring systems

Only systems supplying equipment in these zones are permitted. Metal cable sheaths or metallic covering of wiring systems shall be connected to the supplementary equipotential bonding. Cables should preferably be enclosed in PVC conduit.

Switchgear, control gear and socket outlets

None permitted except for locations where there is no zone 2. In this case, a switch or socket outlet with an insulated cap or cover may be installed beyond 1.25 m from the edge of zone 0 at a height of not less than 300 mm. Additionally, the circuits must be protected by:

1. SELV or
2. Automatic disconnection using a 30 mA RCD or
3. Electrical separation.

Equipment

Only that which is designed for these locations.

Other equipment may be used when the pool/basin is not in use (cleaning, maintenance, etc.) provided the circuits are protected by:

1. SELV or
2. Automatic disconnection using a 30 mA RCD or
3. Electrical separation.

Socket outlets and control devices should have a warning notice indicating to the user that they should not be used unless the location is unoccupied by persons.

Zone 2 (there is no zone 2 for fountains)

Switchgear and control gear

Socket outlets and switches, provided they are protected by:

1. SELV or
2. Automatic disconnection using a 30 mA RCD or
3. Electrical separation.

IP rating of enclosures

Zone 0 IPX8 (submersion)

Zone 1 IPX4 (splashproof) or IPX5 (where water jets are used for cleaning)

Zone 2 IPX2 (drip proof) indoor pools

    IPX4 (splashproof) outdoor pools

    IPX5 (where water jets are used for cleaning).

Supplementary bonding

All extraneous conductive parts in zones 0, 1 and 2 must be connected by supplementary bonding conductors to the protective conductors of exposed conductive parts in these zones.

BS 7671 SECTION 703: HOT AIR SAUNAS

Once again a zonal system, that is, 1, 2 and 3, has been used as per Figure 3.4. In this case the zones are based on temperature.

Additional protection

All circuits in the location should have additional protection against shock by 30 mA RCDs except sauna heater circuits unless recommended by the manufacturer.

Sauna room

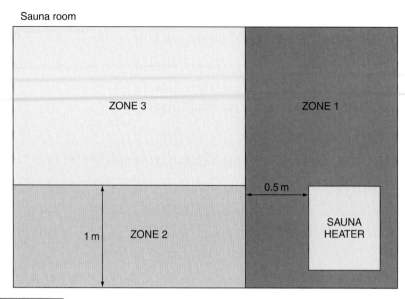

FIGURE 3.4 Sauna room zones.

Wiring systems

It is preferred that the wiring systems for the sauna will be installed outside. However, any wiring inside should be heat resisting and any metal sheaths or conduit must be inaccessible in normal use.

Equipment

All should be at least IPX4 and IPX5 if water jets are to be used for cleaning:

Zone 1 only the sauna equipment
Zone 2 no restriction regarding temperature resistance
Zone 3 must be suitable for 125°C and cable sheaths for 175°C.

Switchgear, control gear and accessories

Only that which is associated with the sauna heater equipment may be installed in zone 2 and in accordance with the manufacturer's instructions. All others should be outside.

BS 7671 SECTION 704: CONSTRUCTION SITES

Not as complicated as one may think. The only areas that require special consideration are where construction work is being carried out, not site huts, etc.

So, let us keep all this as simple as possible. Clearly, construction sites are hazardous areas and in consequence the shock risk is greater.

Protection

For socket outlet circuits of rating up to and including 32 A and circuits supplying equipment of rating up to and including 32 A, the means of protection shall be:

1. Reduced low voltage (preferred for portable hand tools and hand lamps up to 2 kW)
2. Automatic disconnection of supply with additional protection by 30 mA RCDs
3. Electrical separation
4. SELV or PELV (SELV being preferred for portable hand lamps in damp locations).

For socket outlet circuits rated above 32 A, a 500 mA RCD is required.

External influences

These are not addressed in BS 7671, presumably as there are so many different possibilities. So common sense must prevail and equipment used with an appropriate degree of protection in accordance with the severity of the influence.

Wiring systems

Apart from some requirements for flexible cables, the only comment relates to ensuring that cables that pass under site roads, etc. are protected against mechanical damage.

Isolation and switching

An Assembly for Construction Sites (ACS), which is basically the main intake supply board, should comprise a lockable isolator and, for current using equipment and socket outlets:

1. Overcurrent devices
2. Fault protective devices
3. Socket outlets if required.

Plugs and sockets/cable couplers

All should be to BS EN 60309-2.

BS 7671 SECTION 705: AGRICULTURAL AND HORTICULTURAL LOCATIONS

The requirements apply only to locations that do not include the main farmhouse outside of which the environment is hazardous and where, of course, livestock is present (animals are susceptible to lethal shock levels at 25 V AC).

Protection

Protection against shock may be provided by:

1. Automatic disconnection of supply with additional RCD protection for:
   (a) Final circuits supplying socket outlets rated at 32 A or less (30 mA)
   (b) Final circuits supplying socket outlets rated more than 32 A (100 mA)
   (c) All other circuits (300 mA).
2. SELV or PELV.

Protection against thermal effects:

1. Heating appliances should be mounted at appropriate distances from combustible materials and livestock, with radiant heaters at a minimum distance of 0.5 m.
2. For fire protection an RCD rated at 300 mA or less should be used.

Supplementary bonding

Wherever livestock is housed, supplementary bonding must be carried out connecting all exposed and extraneous conductive parts that can be touched by livestock. All metal grids in floors must be connected to the supplementary equipotential bonding.

External influences

1. All equipments must be to at least IP44 and luminaires exposed to dust and moisture ingress, IP54.
2. Appropriate protection for socket outlets where influences are greater than AD4, AE3 and/or AG1.
3. Appropriate protection where corrosive substances are present.

Diagrams

The user of the installation should be provided with plans and diagrams showing the location of all equipment, concealed cable routes, distribution and the equipotential bonding system.

Wiring systems

Any, just as long as it is suitable for the environment and fulfils the required minimum degrees of protection.

A high impact PVC conduit/trunking system would be appropriate in many cases as it is not affected by corrosion, is rodent proof and has no exposed conductive parts. However, the system would be designed to suit the particular environmental conditions.

Wiring systems should be erected so as to be, where possible, inaccessible to livestock. Overhead lines should be insulated and, where vehicles/mobile equipment are used, underground cables should be at least 0.6 m deep and mechanically protected and 1.0 m deep in arable land.

Self-supporting suspended cables should be at a height of at least 6 m.

Switchgear and control gear

Whatever! As long as it is suitable for the conditions and that emergency switching is placed in a position inaccessible to livestock and can be accessed in the event of livestock panic (Stampede!!!).

BS 7671 SECTION 706: CONDUCTIVE LOCATIONS WITH RESTRICTIVE MOVEMENT

These are very rare locations which could include metal tanks, boilers, ventilation ducts, etc., where access is required for maintenance, repair or inspection. Bodily movement will be severely restricted and in consequence such areas are extremely dangerous.

This section deals with the installation inside the location and the requirements for bringing in accessories/equipment from outside.

For fixed equipment in the location, one of the following methods of protection shall be used:

1. Automatic disconnection of supply but with additional supplementary bonding
2. The use of Class II equipment backed up by a 30 mA RCD
3. Electrical separation
4. SELV.

For hand-held lamps and tools and mobile equipment, SELV or electrical separation should be used.

BS 7671 SECTION 708: CARAVAN AND CAMPING PARKS

We drive into a caravan/camping park for our holiday and need to connect to a supply of electricity for all our usual needs. This is accommodated by the provision of suitably placed socket outlets, supplied via distribution circuits.

External influences

Equipment should have at least the following protection codes:

1. IPX4 for the presence of splashes (AD4)
2. IP3X for the presence of small objects (AE2)
3. IK08 for the presence of high severity mechanical stress (AG3) (the IK codes are for impact and 08 is an impact of 5 joules).

Wiring systems

The distribution circuits are erected either underground or overhead:

1. Underground cable (preferred) should be suitably protected against mechanical damage, tent pegs, steel spikes, etc. and at a depth of not less than 0.6 m.
2. If overhead, then 6 m above ground where there is vehicle movement and 3.5 m elsewhere.

Switchgear and socket outlets

1. Supply equipment should be adjacent to, or within 20 m of, the pitch.
2. Socket outlets should be: to BS EN 60309-2; IP44, at between 0.5 m and 1.5 m above ground, rated not less than 16 A and have individual overcurrent and 30 mA RCD protection.
3. If the supply has a PME terminal the protective conductor of each socket needs to be connected to an earth rod.

BS 7671 SECTION 709: MARINAS

This location is basically a camping park for boats and has similar requirements to those of caravan/camping parks.

It is where you arrive in your 40 ft 8 berth cruiser (some hope) looking for a place to park!!!

However, the environment is a little more harsh than the caravan park due to the possibilities of corrosion, mechanical damage, structural movement and flammable fuels, together with the increased risk of electric shock.

External influences

Due to the harsh conditions mentioned the classification of influences would include:

AD water
AE solid foreign bodies
AF corrosion and
AG impact.

Wiring systems

Distribution circuits, like those in caravan parks, can be either underground or overhead, as well as PVC covered mineral insulated, cables in cable management systems, etc.

However, overhead cables on or incorporating a support wire, cables with aluminium conductors or mineral insulated cables shall not be installed above a jetty or pontoon, etc.

Underground cables should have additional mechanical protection and be installed 0.5 m deep.

Overhead cables should be at the same heights as in caravan parks.

Isolation, switching and socket outlets

Generally the same as caravan parks.

Socket outlets should be installed not less than 1 m above the highest water level except that for floating pontoons, walkways, etc. this height may be reduced to 300 mm.

BS 7671 SECTION 710: MEDICAL LOCATIONS

These locations include hospitals, private clinics, dental practices, healthcare centres etc. Installations in such establishments are inevitably complex and hence only a brief overview is considered here.

Rooms in medical locations are divided into groups 0, 1, and 2, Group 0 are massage rooms, group 1 are generally treatment rooms, physiotherapy, hydrotherapy radiology etc. Group 2 are operating and intensive care etc rooms.

Main points:

- Safety sources feeding essential services must be provided i.e. standby systems
- Where required, in groups 1 and 2 only type A or B RCD's are permitted.
- For TN systems, final circuits up to 63A shall have protection by 30 mA or less RCD's.
- IT systems shall be used for equipment and systems in group 2 locations intended for life support and surgical applications, etc.
- Supplementary equipotential bonding shall be provided in each medical location of group 1 and 2.
- Unwanted tripping of 30 mA or less RCD's must be taken into consideration.
- For IT systems in group 2, socket outlets supplying medical equipment must be un-switched and fitted with a supply indicator.
- For initial verification, the additional functional tests of 'insulation monitoring devices' and 'overload monitoring' together with the verification of the integrity of the supplementary bonding must be carried out.

BS 7671 SECTION 711: EXHIBITIONS, SHOWS AND STANDS

This section deals with the protection of the users of temporary structures erected in or out of doors and is typical of antique fairs, motorbike shows, arts and craft exhibitions, etc.

It does not cover public or private events that form part of entertainment activities, which are the subject of BS 7909.

External influences

None particularly specified. Clearly they must be considered and addressed accordingly.

Wiring

Armoured or mechanically protected cables where there is a risk of mechanical damage.

Cables shall have a minimum conductor size of 1.5 mm².

Protection

Against shock:

1. Supply cables to a stand or unit, etc. must be protected at the cable origin by a time-delayed RCD of residual current rating not exceeding 300 mA.
2. All socket outlet circuits not exceeding 32 A and all other final circuits, excepting emergency lighting, shall have additional protection by 30 mA RCDs.
3. Any metallic structural parts accessible from within the unit stand, etc. shall be connected by a main protective bonding conductor to the main earthing terminal of the unit.

Against thermal effects:

1. Clearly in this case all luminaires, spot lights, etc. should be placed in such positions as not to cause a build-up of excessive heat that could result in fire or burns.

Isolation

Every unit, etc. should have a readily accessible and identifiable means of isolation.

Inspection and testing

Tongue in cheek here!! Every installation **should** be inspected and tested on site in accordance with Part 6 of BS 7671.

BS 7671 SECTION 712: SOLAR PHOTOVOLTAIC (PV) SUPPLY SYSTEMS

These are basically solar panels generating DC which is then converted to AC via an invertor. Those dealt with in BS 7671 relate to those systems that are used to 'top up' the normal supply.

There is a need for consideration of the external influences that may affect cabling from the solar units outside to control gear inside.

There must be protection against overcurrent and a provision made for isolation on both the DC and AC sides of the invertor.

As the systems can be used in parallel with or as a switched alternative to the public supply, reference should be made to Chapter 55 of BS 7671.

BS 7671 SECTION 717: MOBILE OR TRANSPORTABLE UNITS

Medical facilities units, mobile workshops, canteens, etc. are the subject of this section. They are self-contained with their own installation and designed to be connected to a supply by, for instance, a plug and socket.

The standard installation protective measures against shock are required with the added condition that the automatic disconnection of the supply should be by means of an RCD.

Also, all socket outlets for the use of equipment outside the unit should have additional protection by 30 mA RCDs.

The supply cable should be HORN-F, oil and flame resistant heavy duty rubber with a minimum copper conductor size of 2.5 mm^2.

Socket outlets outside should be to a minimum of IP44.

BS 7671 SECTION 721: CARAVANS AND MOTOR CARAVANS

These are the little homes that people tow behind their cars or that are motorized, not those that tend to be located on a fixed site. It would be unusual for the general Electrical Contractor to wire new, or even rewire old units. How many of us ever rewire our cars? In consequence, only the very basic requirements are considered here.

Protection

These units are small houses on wheels and subject to the basic requirements of protection against shock and overcurrent. Where automatic disconnection of supply is used this must be provided by a 30 mA RCD.

Wiring systems

The wiring systems should take into account the fact that the structure of the unit is subject to flexible/mechanical stresses and, therefore, our common flat twin and three-core cables should not be used.

Inlets

Unless the caravan demand exceeds 16 A, the inlet should conform to the following:

(a) To BS EN 60309-1 or 2 if interchangeability is required
(b) Not more than 1.8 m above ground level
(c) Readily accessible and in a suitable enclosure outside the caravan
(d) Identified by a notice that details the nominal voltage, frequency and rated current of the unit.

Also, inside the caravan, there should be an isolating switch and a notice detailing the instructions for the connection and disconnection of the electricity supply and the period of time between inspection and testing (3 years).

General

Accessories and luminaires should be arranged such that no damage can occur due to movement, etc.

There should be no compatibility between sockets of low and extra low voltage.

Any accessory exposed to moisture should be IP55 rated (jet proof and dust proof).

BS 7671 SECTION 729: OPERATING AND MAINTENANCE GANGWAYS

Such gangways are likely to be found in restricted areas that are typical of switchrooms, etc., where protection against contact with live electrical parts of equipment is provided by barriers or enclosures or obstacles, the latter having to be under the control of a skilled persons.

Main points:

- Restricted areas must be clearly and visibly marked.
- Access to unauthorized persons is not permitted.
- For closed restricted areas, doors must allow for easy evacuation without the use of a key or tool.
- Gangways must be wide enough for easy access for working and for emergency evacuation.
- Gangways must permit at least a 90° opening of equipment doors etc.
- Gangways longer than 10 m must be accessible from both ends.

BS 7671 SECTION 740: AMUSEMENT DEVICES, FAIRGROUNDS, CIRCUSES, ETC.

This is not an area that is familiar to most installation electricians and hence will only be dealt with very briefly.

The requirements of this section are very similar to those of Section 711 Exhibitions, shows, etc. and parts of Section 706 Agricultural locations (because of animals) regarding supplementary bonding.

For example, additional protection by 30 mA is required for:

1. Lighting circuits, except those that are placed out of arm's reach and not supplied via socket outlets.
2. All socket outlet circuits rated up to 32 A.
3. Mobile equipment supplied by a flexible cable rated up to 32 A.

Automatic disconnection of supply must be by an RCD.

Equipment should be rated to at least IP44.

The installation between the origin and any equipment should be inspected and tested after each assembly on site.

BS 7671 SECTION 753: FLOOR AND CEILING HEATING SYSTEMS

Systems referred to in this section are those used for thermal storage heating or direct heating.

Protection

Against shock:

1. Automatic disconnection of supply with disconnection achieved by 30 mA RCD.
2. Additional protection for Class II equipment by 30 mA RCDs.
3. Heating systems provided without exposed conductive parts shall have a metallic grid of spacing not more than 300 mm installed on site above a floor system or below a ceiling system and connected to the protective conductor of the system.

Against thermal effects:

1. Where skin or footwear may come into contact with floors the temperature shall be limited, for example to 30°C.
2. To protect against overheating of these systems the temperature of any zone should be limited to a maximum of 80°C.

External influences

Minimum of IPX1 for ceilings and IPX7 for floors.

The designer must provide a comprehensive and detailed plan of the installation which should be fixed on or adjacent to the system distribution board.

Questions

1. When is the space under a bath or shower outside of all the zones?
2. What is the IP rating for equipment in zone 1 of a swimming pool where jets are use for cleaning purposes?
3. What is the maximum height above floor level of zone 2 of a sauna?
4. What rating of RCD id required for a 63 A socket outlet on a construction site?
5. How deep should a cable be buried in arable ground on a farm?
6. What is the maximum rating of an RCD used in conjunction with class II fixed equipment in a conductive location with restricted movement?
7. What is the impact code for equipment subject to (AG 3) mechanical stress on a caravan park?
8. What is the minimum height of a socket outlet above the highest water level for a walk-way in a marina?
9. What is the room Group reference for an operating theatre in a hospital?
10. What rating and type of RCD should be provided at the origin of the supply cable to a stand in an exhibition?
11. For a PV installation, where should inverter isolation be provided?
12. Which socket outlets associated with mobile units should be 30 mA, or less, protected?
13. What is the maximum height above ground level for the inlet to a caravan?
14. Above what distance should a gangway be accessible from both ends?
15. What is the minimum IP rating for equipment in a fairground?
16. What is the minimum IP rating for ceiling heating systems?

Answers

1. When accessible only by a tool
2. IPX5
3. 1 metre
4. 500 mA
5. 1 metre
6. 30 mA or less
7. IK08
8. 300 mm
9. Group 2
10. 300 mA time delayed or 'S' type
11. on both the a.c. and the d.c. side
12. Those used for equipment outside the unit
13. 1.8 m
14. 10 m
15. IP44
16. IPX1

BS 7671 Appendices

There are 14 appendices in BS 7671, only one of which is a requirement, that is number 1; all the rest are for information.

1. British Standards referred to in BS 7671
2. A list of Statutory Regulations, etc.
3. Time/current curve for protective devices
4. Tables of cable current-carrying capacities and voltage drops
5. Classification of external influences
6. Certificates and schedules
7. Harmonized cable core colours
8. Tables of cable current-carrying capacities and voltage drops for busbar and powertrack systems
9. Definitions – multiple source, DC and other systems
10. Protection of conductor's in parallel against overcurrent
11. Not used
12. Not used
13. Measurement of resistance of non-conducting floors and walls
14. Earth fault loop impedances using the 0.8 factor
15. Arrangement of ring and radial final circuits.
16. Devices for protection against over current.

Sample Questions

This appendix looks at typical C&G 2392-30 examination questions and Appendix 3 gives suggested solutions expected by the examiners. Clearly, in many instances there is not always one correct answer, and the examiner will have a range of alternatives from which to award marks. Owing to the time constraints, approximately 18 minutes per question, the candidate is not expected to, and nor can he or she, write an essay in answer to descriptive questions. All that is required are reasoned statements which indicate a knowledge and understanding of the subject matter, and if time allows, specific reference to relevant parts of the 17th Edition, although this is not essential unless asked for. It is not sufficient to simply quote Regulation numbers or Parts in answer to a question. In fact, no marks are awarded for such answers.

Example A2.1

A factory manufacturing chemical products is situated close to the supply transformer feeding an industrial estate. The earthing system is TN-C-S with a measured loop impedance of $0.015\,\Omega$ and PFC of 16 kA. It is required to increase the level of lighting by installing 26 400 W/230 V high-bay discharge luminaires. The existing wiring system is a mixture of PVC/SWA cables and galvanized trunking and conduit. There is no spare capacity in any of the existing distribution fuse boards.

Outline the design considerations for the new lighting, with regards to:

1. Maximum demand and diversity.
2. Maintainability.
3. External influences.
4. Wiring system.
5. Control and protective devices.

Example A2.2

A consumer is having major alterations to their premises, one part of which is to convert an existing kitchen extension to a pottery room housing a 9 kW/230 V fan-assisted kiln and it is proposed to utilize the existing cooker circuit to supply it. The cabling is 6.0 mm² twin with 2.5 mm² cpc, clipped direct throughout its 25 m run and protected by a 32 A BS EN 60898 Type B CB and there are no adverse conditions prevailing. The external value of loop impedance has been measured as 0.3 Ω. Show by calculation what changes, if any, are required to enable the existing system to be used.

Example A2.3

You are to provide the temporary electrical installation for a construction site on which the site huts and offices together with the main supply point are on the opposite side of the access road to the building under construction. The services required are supplies for:

1. The site huts and offices.
2. Portable tools
   (a) Indicate a suitable method of running supplies from the site hut area to the construction area. What type of sockets and cable couplers should be used?
   (b) State the voltages and disconnection times for 1 and 2 above.
   (c) If one of the circuits for the portable tools is protected by a 16 A Type B CB, what is the maximum value of the loop impedance Z_s for that circuit?

Example A2.4

Part of a farm complex supplied by a TT system is to be converted for use as a poultry incubation area. The existing wiring is some 30 years old and incorporates a voltage-operated earth leakage circuit breaker (ELCB). Outline the design criteria to be considered with regards to:

1. The wiring system.

2. Protection against shock.
3. Protection against thermal effects.

Example A2.5

A single-phase distribution circuit to a distribution board housing BS 88 fuses is wired in 6.0 mm² SWA/XLPE cable.

A radial lighting circuit wired in 1.5 mm² PVC copper cable with a 1.5 mm² cpc and protected by a 10 A BS 88 fuse is fed from the board. The length of the lighting circuit is 40 m.

The measured value of Z_s at the distribution board is 2.1 Ω, and the ambient temperature at the time of measurement was 20 °C.

(a) What would be the minimum gross size of the distribution cable armouring if it is to be used as the cpc?
(b) Calculate the value of Z_s at the extremity of the lighting circuit.

Is this value acceptable?

Example A2.6

During a periodic test and inspection of the installation in a butcher's shop, it is revealed that the circuit supplying an electrically operated compressor does not meet the maximum earth fault loop impedance requirements. The circuit is protected by a 16 A Type C CB, and the unit is situated 1 m from a steel sink. Explain how, under certain conditions, this situation may be resolved by the use of supplementary bonding. Support your answer with calculation.

Example A2.7

A 2.5 mm² ring final circuit 60 m long is wired in singles in a PVC conduit; the cpc is 1.5 mm². A ring circuit continuity test is performed involving measurements at each socket.

1. What is the purpose of this test?
2. Explain a method of identifying the opposite 'legs' of the ring.

3. What would be the reading between L and E at the socket nearest to the mid-point of the ring?
4. What is the significance of this mid-point reading?

Example A2.8

1. Give three Examples for the use of an RCD, indicating residual operating currents and operating times.
2. How often should a consumer operate an RCD via its test button? What does this test achieve?
3. Give one Example for the use of a time-delayed RCD.

Example A2.9

Figure A2.1 shows a ring final circuit wired in flat twin with cpc cable $2.5\,\text{mm}^2 + 1.5\,\text{mm}^2$. The protection is by 32 A Type B CB. If a test for continuity was performed at $15\,°C$ using the measurement at each socket method, calculate:

1. The reading at each socket between L and N.
2. The value of $R_1 + R_2$.
3. The value of Z_s for comparison with the tabulated maximum value.

Is this value acceptable?
($2.5\,\text{mm}^2$ copper has a resistance of $7.41\,\text{m}\Omega/\text{m}$, $1.5\,\text{mm}^2$ is $1.21\,\text{m}\Omega/\text{m}$ and the value of Z_s is measured as $0.28\,\Omega$.)

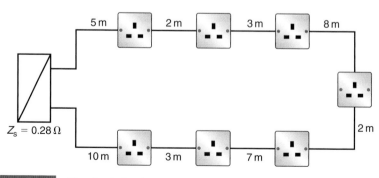

FIGURE A2.1   Ring final circuit.

Example A2.10

A small three-storey commercial office complex is due to have a periodic test and inspection. Outline the major steps you would take regarding:

1. Disturbance to office routine.
2. Meeting the requirements of the Electricity at Work Regulations 1989.
3. Measuring the continuity of main protective bonding conductors.
4. Reporting defects and issuing certificates.

Example A2.11

1. Give two reasons, when conducting an insulation resistance test on a large complex installation, for breaking it down into smaller sections. What precautions should be taken before commencing the tests?
2. The test results for each section of such an installation are $50\,M\Omega$, $20\,M\Omega$, $100\,M\Omega$ and $4\,M\Omega$. Show by calculation the expected overall insulation resistance at the intake position.

Example A2.12

1. A wiring system employing the use of singles in steel trunking is to be installed. Outline the main design and installation considerations with regards to this installation.
2. The trunking at one point will accommodate the following single-stranded conductors:

28	$1.5\,mm^2$
20	$2.5\,mm^2$
12	$6.0\,mm^2$
10	$10.0\,mm^2$

Determine the minimum size of trunking to be used.

Suggested Solutions to Sample Questions

Example A3.1

1. Determine the new maximum demand by calculating the increase in load and adding to the existing maximum demand. Check that suppliers' equipment and main switchgear/busbars, etc., can accommodate the extra load.

$$\text{Increase in load} = \frac{\text{power} \times (1.8 \text{ for discharge lamps})}{230}$$

$$= \frac{26 \times 400 \times 1.8}{230} = 81.4 \text{ A}$$

No diversity would be allowed as it is likely that the lamps will be on all the time.

2. Luminaires need to be accessible for cleaning, repair, lamp replacement, etc.
   - Access equipment should be available
   - Spare lamps, chokes, etc., should be kept
   - Luminaires supplied via plug and socket arrangement to facilitate easy removal, and without losing supply to the other lamps.

3. As chemicals are being produced the atmosphere could be corrosive, and there may be a fire risk; external influences classification would be AF2 and BE2.

4. If valid documentation exists it is possible for a decision to be made to use, at least in part, the existing trunking system. If not, and this is most likely, a new system should be installed using either singles in galvanized steel trunking and conduit, or PVC sheathed material insulated cable, or PVC sheathed SWA cable, with circuits spread over three phases. Protect against shock by automatic disconnection of supply.

5. Control by switch operating a three-phase 80 A contactor feeding a three-phase distribution board housing BS 88 fuses to cater for

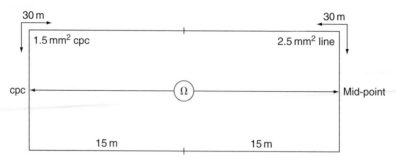

FIGURE A3.1 Ring with 2.5 mm² line and 1.5 mm² cpc.

Table A3.1 Resistance of Copper Conductors at 20 °C

Conductor CSA (mm²)	Resistance (mΩ/m)
1.0	18.1
1.5	12.1
2.5	7.41
4.0	4.61
6.0	3.08
10.0	1.83
16.0	1.15
25.0	0.727
35.0	0.524

the high PFC at the intake. Contactor and DB must also be able to handle the PFC.

Example A3.2

$$\text{Kiln design current} = \frac{P}{V} = \frac{9000}{230} = 39 \text{ A}$$

CB setting I_n, such that $I_n \geq I_b = 40 \text{ A}$

No correction factors, hence $I_t \geq 40 \text{ A}$

$$I_t = 40 \text{ A}$$

Cable size $= 6.0 \text{ mm}^2$

$$\text{Volt drop} = \frac{\text{mV} \times I_b \times L}{1000} = \frac{73 \times 39 \times 25}{1000} = 71.18\,\text{V ok}$$

$$\text{Shock risk } Z_s = Z_e + (R_1 + R_2) = 0.3 + \frac{(3.08 + 7.41) \times 25 \times 1.2}{1000}$$

$$= 0.615\,\Omega \text{ ok, as } Z_s \text{ maximum is } 1.44\,\Omega$$

$$\text{Thermal constraints } I = \frac{U_0}{Z_s} = \frac{230}{0.615} = 374\,\text{A}$$

$$t < 0.1\text{s for a } 40\,\text{A Type B CB}$$

$$k = 115$$

$$S = \frac{\sqrt{I^2 t}}{k} = 1.03\,\text{mm}^2$$

So the 2.5 mm² cpc is ok. The only change to the existing installation would be to uprate the 32 A CB to 40 A.

Example A3.3

(a) PVC sheathed SWA cable supported over the access road on a catenary wire at a minimum height of 5.8 m. Cable couplers and sockets should be to BS EN 60309-2.

(b) Site huts and offices   Portable tools
      230 V   0.4 s and 5 s   110 V CTE (preferred) 5 s

(c) From Table 41.6, IEE Regulations, $Z_s = 6.9\,\Omega$ or, calculate from:

$$Z = \frac{U_0}{I_a}$$

I_a for a 16 A Type B CB from Figure 3.4 in IEE Regulation curves = 80 A

So $Z_s = 55/80 = 0.687\,\Omega$

Example A3.4

1. It is unlikely that the existing wiring will meet the new requirements, and due to its age it would be best to replace with a new

all-insulated system, for example singles in PVC conduit out of reach of livestock and supplied by a manufacturer who specifies resistance to the onerous conditions found on farms.

2. Remove the voltage-operated ELCB. These are not permitted. RCD protection will be needed for: socket outlet circuits up to 32A (30 mA), over 32A (100 mA) and all other final circuits (300 mA). Protection against shock would be by automatic disconnection of supply with supplementary equipotential bonding.

3. Protection against fire may be achieved by using an RCD rated up to 300 mA, except where equipment essential to the welfare of livestock is involved.

Incubation and subsequent hatching involve the use of infrared lamps to maintain a stable temperature. The enclosures of such lamps may become hot and hence must be located in positions that will not cause fire or burns. Radiant heaters should not be located less than 0.5 m from livestock or combustible materials.

Example A3.5

(a) Referring to Table 54.7:

$$\text{Sp} = \frac{k1S}{k2} \, (\text{Sp is the CSA of the protective conductor})$$

As XLPE is thermosetting:

$$k1 = 143 \text{ and } k2 = 46$$

$$\text{Sp} = \frac{143 \times 6}{46} = 18.6 \, \text{mm}^2$$

(b) Distribution circuit $Z_s = 2.1 \, \Omega$.

$$\text{Final circuit} \, (R_1 + R_2) = \frac{(12.1 + 12.1) \times 40 \times 1.2}{1000} = 1.16 \, \Omega$$

So total $Z_s = 2.1 + 1.16 = 3.26 \, \Omega$

Ok, as Z_s maximum for a 10A BS 88 fuse for a final circuit not exceeding 32A is 5.11 Ω.

Example A3.6

Provided that the value of loop impedance allows a fast enough discon-nection time to protect against thermal effects, then compliance with the Regulations may be achieved by connecting a supplementary bond-ing conductor between the exposed conductive parts of the compressor and the sink. The resistance of such a conductor must be less than or equal to:

$$\frac{50}{I_a}$$

I_a, the current causing operation of the protection within $5\,\text{s}$, is $160\,\text{A}$ for a Type C CB.

$$\text{So } R \leq \frac{50}{160} = 0.313\,\Omega$$

Example A3.7

1. To identify breaks in the ring and/or interconnections across the ring.
2. Test with a low reading ohmmeter between each L, N and E leg and the corresponding terminal at the nearest socket. A low value indicates the short leg, a high value the long leg.
3. Reading at mid-point (Fig. A3.1 and table A3.1)

$$\text{Reading at mid-point} = \frac{30 \text{ m of } 2.5 \text{ mm}^2 + 30 \text{ m of } 1.5 \text{ mm}^2}{2}$$

$$= \frac{30 \times 0.00741 + 30 \times 0.0121}{2} = 0.293\ \Omega$$

4. This value is $(R_1 + R_2)$ for the ring.

Example A3.8

1. (a) When the loop impedance value for an overcurrent device cannot be met. The product of the residual operating current

of the device and the loop impedance should not exceed 50 V. The device should trip within 200 ms at the rated residual current.

(b) If additional protection against shock is required. RCD should be rated at 30 mA or less, and trip within 40 ms at a residual operating current of 150 mA.

(c) In agricultural situations, for protection against fire. The RCD should not be rated above 300 mA, and used for circuits other than those essential for the welfare of livestock. The tripping time would be within 200 ms at a rated residual current.

2. The RCD should be tested quarterly via the test button. This only checks the operating mechanism not any earthing arrangements.

3. On a TT system where the whole installation is protected by, say, a 100 mA device and the sockets by a 30 mA device. A time delay on the 100 mA RCD will give discrimination with the 30 mA RCD.

Example A3.9

1. The ring is 40 m long, so the L to N reading at each socket would be:

$$\frac{2 \times 40 \times 7.41}{1000 \times 4} = 0.148 \ \Omega$$

2. The mid-point $R_1 + R_2$ is 20 m of 2.5 mm

$$+20 \text{ m of } 1.5 \text{ mm in parallel} : = \frac{20 \times (7.41 + 12.1)}{1000 \times 2} = 0.195 \ \Omega$$

3. $(R_1 + R_2)$ corrected for $15\,°C = 0.195 \times 1.02 = 0.199 \,\Omega$

Correction for operating temperature $= 0.199 \times 1.2 = 0.239 \,\Omega$

So total $Z_s = 0.28 + 0.239 = 0.52 \,\Omega$. This is ok as tabulated maximum value is $1.44 \,\Omega$.

Example A3.10

1. Careful planning and consultation will be required before any work commences. It may be the case that access to the premises is better suited to a weekend or evenings when no staff are present. If this is not possible, then the installation should be tested in small sections, all the tests required in each section being done at that time. Clearly, in the modern office, computers play a major role, and unless UPS are present, advice should be sought before isolating any supplies.

2. The inspector is a duty holder and as such must take all precautions to safeguard himself and others. Visual inspection can involve entry into enclosures housing live parts, and unless it is completely impracticable, supplies must be isolated and locked off. Testing on or near live equipment is prohibited unless it is unreasonable for it to be dead, for example loop impedance and RCD tests. All test equipment must be suitable for the use intended and should be in a safe condition. All test results must be recorded.

3. It is usual for the bonding conductor to be disconnected for test purposes. Unless all supplies to the complete installation can be isolated, bonding conductors must not be disconnected.

4. All test results and details of the inspection must be entered on to schedules and a periodic report given to the person ordering the work. The report should include details of the extent of the work, any dangerous situations prevailing, restrictions to the inspection and test, and serious defects.

Any certificate issued should indicate and explain departures from the 17th Edition, especially those due to installations constructed before the current Regulations.

Example A3.11

1. Large installations may have circuits in parallel which can result in pessimistically low values of insulation resistance even though there are no defects. Dividing the installation into smaller sections will overcome these low readings.

Subdivision of the installation, especially on periodic inspections, will enable minimum disruption of work processes.

All persons must be informed that the test is to take place, all supplies isolated from the part of the installation in question, all electronic devices, capacitors, neon indicators, etc., should be disconnected, and ensure that no electrical connection exists between any live conductor and earth.

2. The overall value will be the sum of the individual insulation resistances in parallel, hence:

$$\frac{1}{R} = \frac{1}{R_1} + \frac{1}{R_2} + \frac{1}{R_3} + \frac{1}{R_4} = \frac{1}{50} + \frac{1}{20} + \frac{1}{100} + \frac{1}{4}$$

$$= 0.02 + 0.05 + 0.01 + 0.25 = 0.33$$

$$\frac{1}{R} = \frac{1}{0.33} = 3\,\text{M}\Omega$$

Example A3.12

1. The design should embrace grouping of circuits, space factor if trunking sizes are outside the scope of tabulated sizes, and external influences which may affect the choice of trunking finish.

With regard to the installation, supports must be at the correct spacing, joints should be bridged with an earth strap, and where trunking passes through walls, ceilings, etc., it should be externally and internally sealed to the level of the fire resistance required for the building construction.

2. Using the tabulated conductor and trunking terms, we have:

$$28 \times 8.1 = 226.8$$
$$20 \times 11.4 = 228$$
$$12 \times 22.9 = 274.8$$
$$10 \times 36.3 = 363$$
$$\text{Total} = 1092.6$$

Hence trunking size is $75 \times 375\,\text{mm}^2$.

For information

Index